Mobility in the Face of Extreme Hydrometeorological Events 1

Series Editor
Françoise Gaill

Mobility in the Face of Extreme Hydrometeorological Events 1

Defining the Relevant Scales of Analysis

Edited by

Céline Lutoff
Séverine Durand

ISTE PRESS

ELSEVIER

First published 2018 in Great Britain and the United States by ISTE Press Ltd and Elsevier Ltd

ISTE Press Ltd
27-37 St George's Road
London SW19 4EU
UK

www.iste.co.uk

Elsevier Ltd
The Boulevard, Langford Lane
Kidlington, Oxford, OX5 1GB
UK

www.elsevier.com

British Library Cataloguing-in-Publication Data
A CIP record for this book is available from the British Library
Library of Congress Cataloging in Publication Data
A catalog record for this book is available from the Library of Congress
ISBN 978-1-78548-289-2

Contents

Preface

Flash floods often get people talking, especially in the South of France and in the whole of the Mediterranean Basin. In a context of climate change, we must acknowledge that these violent phenomena are increasing [ROC 12] and affecting previously spared areas, as was the case in the North of France [DOU 08]. Understanding the dynamics of these phenomena is essential for learning how to become protected and to mitigate damage. But focusing solely on the physical phenomenon is not enough. In order to reduce potential damage, especially in terms of human lives, it is necessary to understand how these hydrometeorological dynamics interact with social dynamics, build up and lead to a catastrophic event.

Research carried out on the subject has highlighted the central role of mobilities in this interaction. In an article from 1983, French *et al.* identified daily mobility as one of the major causes of death during flash floods [FRE 83]. According to systematic reports by the National Weather Service (NWS reports), motorists accounted for 43% of flash flood victims in the United States between 1969 and 1981. This observation was confirmed by other studies carried out in the United States [MOO 83] or in other locations, more recently [ANT 01, COA 99, JON 05a]. A detailed analysis shows that mobility-related casualties are more specific to certain categories of people and types of watersheds [RUI 08, RUI 09, TER 17]. During the 2002 event in the Gard department, in southern France, Ruin *et al.* found that from among the 22 people who lost their lives, half were outside, and their profile was clearly different from the victims who died at home [RUI 08]. Accidents involving a situation of mobility took place in small watersheds (less than 20 km^2), characteristic of very short reaction times. They mostly involved middle-aged men (43 years old on average), while victims found at home were older (76 years old, on average) and lived in larger watersheds (over 1,000 km^2), involving slower flood dynamics.

Chapter written by Céline LUTOFF and Séverine DURAND.

It seems, then, that the dynamics of the small watercourses combined with the particular dynamics of mobility is regularly conducive to catastrophe. Harnessing these dramatic situations demands an understanding of what is at stake in this particular interaction between the dynamics of the physical phenomenon and social dynamics when the event ultimately takes place. Our hypothesis is that grasping the notion of scale plays a key role in this articulation.

In order to better understand these processes, an interdisciplinary team bringing together specialists in hydrology, meteorology, humanities and the social sciences from different organizations[1] has undertaken several successive research projects, namely with the support of the *Agence Nationale de la Recherche française* (ANR). Among these, the *MobiClimEx* project was funded by the ANR *Société & Environnement* in 2012 and became part of the *HYMEX* international program (HYdrological Cycle in the Mediterranean Experiment, www.hymex.org), dealing with the evolution of the water cycle in the Mediterranean Basin in a context of climate change. This last project focuses on the evolution of mobility exposure in the face of the flash floods that have taken place in the south east of France since the 1960s, and more specifically, in the Gard department, which is regularly exposed to this type of event. The project made it possible to consolidate previously acquired knowledge about this area, and more largely about the Mediterranean area, which we intend to gather in this book.

In parallel, a reflection on the question of scales has been carried out since 2010 at Grenoble's university center, within the framework of an interdisciplinary seminar called *AMETIhST* (www.ametihst.fr). This seminar intends to build the necessary interdisciplinarity for addressing the complex issues raised here, but more generally, it seeks to better understand the interactions between society and the environment. Taking place once a month, the AMETIhST seminar brings together geoscientists and human and social sciences researchers from the Grenoble area. The aim is to share a common bibliographic culture explored through the lenses of particular theme: the interaction between society and the environment. We have collectively exploited various references dealing with the question of scales and mobilized these reflections so as to better understand the connections between mobilities and flash floods within the framework of the *MobiClimEx* project.

Thus, this interdisciplinary work brings forward several concepts that we have taken care to define and that we want to share as an introduction to this book in order

1 The interdisciplinary team has been built progressively since its beginnings in the 2000s, gathering researchers from different laboratories, such as the LTHE (which later became IGE) and PACTE in Grenoble, the CNRM from *Météo France* in Toulouse and the IRSTEA center from Lyon-Villeurbanne.

to help the reader understand the meaning that we attribute to some of the observations made herein. The concept of scale is evidently one of the first to elicit questions and which requires a common definition. In fact, to speak of small or large scale does not entail the same consequences for a geographer or a hydrometeorologist or a sociologist. Many geographers consider scale in its cartographic definition, rather than as a geographic scale (as Racine *et al.* observed [RAC 80]). In this case, the "small scale" corresponds to the vision of a very large space (in accordance with the cartographic scale corresponding to a planisphere), while the "large scale" represents a small space. Contrary to this, in other disciplines, hydrometeorology in particular, the term scale is employed in its most commonly understood sense: a small scale refers to a restricted space, as seen through a magnifying glass, while the large scale corresponds to a more global vision. In order to make the reading of this book more accessible to a broader audience, encompassing all disciplines and specialties, it is this common definition that we will employ here.

Following the same intention to facilitate the understanding of the chapters that will follow, here we propose clarifying some of the other central concepts of this book, some of which will be largely redefined and/or discussed later. Our interest particularly lies in extreme hydrometeorological phenomena, specifically in what concerns flash floods. These phenomena share certain characteristics that will be explained in Chapter 3, but here we can state that these are sudden floods (which can be triggered in a few minutes and last only for a few hours), violent (the recorded flows often reach records, and some witnesses have referred to walls of water in order to describe what they saw), particularly destructive (since they are among the deadliest natural phenomena, with much higher mortality rates than slow floods [JON 05b]), and those especially affecting social mobility, as many studies have shown.

In this work, mobilities are understood through different and complementary time scales: on the one hand, daily mobility, and, on the other hand, residential mobility. Daily mobility concerns everyday commuting from home to various activities (professional, social, leisure, etc.). These have been a recurring object of study since the 1960s in the social sciences, and particularly in economics, urban planning and geography [BRU 93, BON 04]. Residential mobility, formerly known as migrations [BRU 93], instead refer to the movement of populations throughout a life trajectory, as a result of changing residence and varying career paths, specifically professional ones. These movements are particularly studied by demographers, economists and geographers, on the basis of recurring data collected by certain national organizations such as INED or INSEE in France [BRU 93], allowing both a broader vision of these vital trajectories as well as their impact on spaces. For our study, their interest lies in the interactions they establish with the dynamics of hydrometeorological phenomena: how flash flood events are envisioned within the residential area, be it at an individual level (the decision of living near a river) or at a community level (through land-use management in flood-prone areas).

The concept of interactions that we have outlined since the beginning of this introduction follows the definition offered by the *Centre National de Resources Textuelles et Lexicales* (http://www.cnrtl.fr) as "a reciprocal action [...] between two or more phenomena". So, the idea here is to consider the reciprocal action of extreme hydrometeorological phenomena (floods) and the phenomenon of social mobility. This involves not only exploring how flash floods may have an impact on daily and residential mobility on the one hand, but also studying how these mobilities can reduce or, on the contrary, worsen flood phenomena or their impact (particularly by modifying land use). This notion of impact will often be quoted within the framework of this work, since it constitutes one of the clearest manifestations of the interactions that we wish to explore.

In Chapter 1, we will explore useful scales for better apprehending these interactions. Chapter 2 will deal with the social effects of flash floods using concrete examples, collected after flash floods, through the exploration of interviews with the residents who have witnessed the events or who live in flood-prone areas. In Chapter 3, we will discuss the specifics of flash floods and the difficulties in forecasting them. Chapter 4 will question the notion of impact, as a manifestation of the interactions between floods and mobility. The chapter will aim to define the characteristics of high-impact events and the necessary conditions for their eventual prediction. In Chapter 5, we will discuss possible regulatory impacts through a historical overview of how different institutional scales are set in motion after major floods. Chapter 6 will focus on how these regulations are implemented at a local level as well as the barriers and constraints that they face. Chapter 7 will examine the major migrations that took place in the south of France from the 1960s onwards, as well as their possible connection with major hydrometeorological events, particularly resulting from land use evolution. Finally, Chapter 8 will aim to synthesize the various contributions of scale analysis and to open research perspectives on these situations involving a conflict between social mobility and water mobility.

Before going into the heart of the matter, we would like to thank all the people and the structures that contributed to the development of this book and its publication. First of all, we would like to thank the *Agence Nationale de la Recherche*, and more specifically the research program *Société et Environnement*, which financially supported the *MobiClimEx project* (under decision no. ANR-12-SENV-0002), and contributed to the development of much of the work presented in this book. We would also like to thank ISTE for placing their trust in us and for helping us to materialize this publication. Special thanks are devoted to the authors who not only contributed during the writing process, but also when proofreading chapters of other authors. In order to guarantee the best possible understanding of this book, which clearly intends to be interdisciplinary, we asked the authors to critically reread some chapters which were distant from their original discipline. The idea of these proofreadings was not only to contribute to sharing the knowledge acquired within each of the disciplines, but also to ensure that the different dimensions of the issue of mobility in the face of extreme natural phenomena could be understood by everyone. We would also like to thank Hervé Gumuchian, geographer, former professor at Joseph Fourier Grenoble University who, although not a specialist in these issues, agreed to a general reading of the book in order to guarantee its overall coherence and readability. Our thanks also go to all the people who gave us a share of their time during our field surveys in order to comment on their experiences with flash floods and living in flood-prone areas. As a final point, the reflections presented in this book owe much to the exchanges we had in the context of the AMETIhST interdisciplinary seminar (www.ametihst.fr). Thanks to its members, who fully contributed to the reflection about scales throughout 2016, and thanks to Michel Grossetti, who agreed to bring his expert eye to this collective work. We hope that the reader will enjoy reading this book as much as we enjoyed working together to write it.

Bibliography

[ANT 01] ANTOINE J.-M., DESAILLY B., GAZELLE F., "Les crues meurtrières, du Roussillon aux Cévennes", *Annales de Géographie*, vol. 622, pp. 597–623, 2001.

[BON 04] BONERANDI E., "De la mobilité en Géographie", *Géoconfluences*, 2004.

[BRU 93] BRUN J., "La mobilité résidentielle et les sciences sociales: transfert de concept et questions de méthodes", *Les Annales de la recherche urbaine*, vol. 59, no. 1, pp. 3–14, 1993.

[COA 99] COATES L., "Flood fatalities in Australia, 1788–1996", *Australian Geographer*, vol. 30, no. 3, pp. 391–408, 1999.

[DOU 08] DOUVINET J., Les bassins versants sensibles aux 'crues rapides' dans le Bassin Parisien-Analyse de la structure et de la dynamique de systèmes spatiaux complexes, Doctoral dissertation, University of Caen Normandy, 2008.

[FRE 83] FRENCH J., ING R., VON ALLMEN S. *et al.*, "Mortality from flash floods: a review of national weather service reports, 1969–81", *Public Health Reports*, vol. 98, no. 6, p. 584, 1983.

[JON 05a] JONKMAN S.N., KELMAN I., "An analysis of the causes and circumstances of flood disaster deaths", *Disasters*, vol. 29, no. 1, pp. 75–97, 2005.

[JON 05b] JONKMAN S., "Global perspectives on loss of human life caused by floods", *Natural Hazards*, vol. 34, pp. 151–175, 2005.

[MOO 83] MOONEY L., "Applications and implications of fatality statistics to the flash flood problems", *Proceedings of the 5th Conference on Hydrometeorology*, pp. 127–129, 1983.

[RAC 80] RACINE J.B., RAFFESTIN C., RUFFY V., "Echelle et action, contributions à une interprétation du mécanisme de l'échelle dans la pratique de la géographie", *Geographica Helvetica*, vol. 35, no. 5, pp. 87–94, 1980.

[ROC 12] ROCHE P.-A., MIQUEL J., GAUME E., *Hydrologie quantitative: Processus, modèles et aide à la décision*, Springer Science & Business Media, 2012.

[RUI 08] RUIN I., CREUTIN J.-D., ANQUETIN S. *et al.*, "Human exposure to flash floods – relation between flood parameters and human vulnerability during a storm of September 2002 in Southern France", *Journal of Hydrology*, vol. 361, nos 1–2, pp. 199–213, 2008.

[RUI 09] RUIN I., CREUTIN J.-D., ANQUETIN S. *et al.*, "Human vulnerability to flash floods: Addressing physical exposure and behavioral questions", *Flood Risk Management: Research and Practice Proceedings of the European Conference on Flood Risk Management Research into Practice (FLOODrisk 2008)*, Oxford, UK, September 30–October 2, 2008, pp. 1005–1012, 2009.

[TER 17] TERTI G., RUIN I., ANQUETIN S. *et al.*, "A situation-based analysis of flash flood fatalities in the United States", *Bulletin of the American Meteorological Society*, vol. 98, no. 2, pp. 333–345, 2017.

Chapter Summaries

In order to have a quick overview of the content of each chapter, here we provide a summary of each one in just a few lines.

Chapter 1: Mobility Exposure Scales of Analysis in the Face of Flash Floods

This chapter aims to explore the pertinence of the notion of scales in the analysis of mobility exposure to flash floods. By mobilities, we mean both daily mobility and residential mobility. Daily mobilities have been identified in the literature as particularly vulnerable to flash floods, although an assessment of exposure has been developed very briefly. As for the exposure of residential mobility, it addresses the possible recurrence of extreme hydrometeorological phenomena in a context of climate change. How can we grasp the interaction between these two types of social processes and the dynamics of flash flood phenomena? Here, we put forward the hypothesis that the notion of scale is central to a better understanding of what is at play when catastrophic events occur. We explore the contributions of this notion, as seen through the human–environment interaction approach and in the analysis of social processes. This analysis leads us to suggest a synthetic scale expressing the spatial and temporal constraints that flash floods impose on social dynamics.

Chapter 2: Social Effects of Floods on Mobilities

This chapter focuses on what the breakdown situation of a flash flood creates from a social point of view. This questioning addresses both the short time of the event – during which the aim is to understand how inhabitants react at the moment of the crisis – and over longer time scales, in which the aim is to understand how the idea of flooding influences residential choices and the ways of living. We therefore articulate our analyses over two timescales: that of human decisions, collected through narratives of the actions undertaken during the crisis (short timeframes), and that of residential trajectories, collected using the method of life narratives (long timeframes). This is done in order to highlight the forms of convergence in the social logic uncovered at these two timescales, and notably that of the "power of the daily routine".

Chapter 3: The Challenges of Flash Flood Forecasting

This chapter introduces the characteristic spatial and temporal scales of flash floods and associated hydrometeorological processes. These scales impose a specific strategy of observation and modeling to better understand and predict the forcing of pluviometry and the hydrological responses of the impacted catchment basins. Further on in the chapter, after a presentation of the components of a flood prediction system, a state-of-the-art flash flood prediction model shows the significant progress made over the recent period due in particular to the availability of pluviometry information from meteorological radars to the advances in meteorological prediction and the development of distributed hydrological models adapted to the flash flood scales. Finally, an example of a predicted result for a recent forecasted event illustrates the potential of these tools for better predicting such episodes and the perspectives for development to head toward the prediction of the impact of these phenomena.

Chapter 4: High-impact Weather Events: Is a Socio-hydrometeorological Characterization Possible?

This chapter questions the notion of a high-impact event and suggests several ideas for the establishment of a more objective definition. To do so, a more thorough knowledge of the interrelations between hazards and the associated socioeconomic impacts is useful and requires multidisciplinary observations to describe and understand these links, related to the affected populations and their anticipation and reaction capacity. By targeting these reflections on flash flood events in the south-east of France, the cross-analysis of different national and international database showed a great heterogeneity of available data, thus making such interdisciplinary studies difficult. We therefore propose to guide the definition of a high-impact event through the resonance of spatial and temporal scales associated with hazards and impacts by mobilizing pluviometry, socioeconomic and media data.

Chapter 5: Legal Evolution of the Confrontation Between Water Mobility and the Mobility of People

At an institutional level, the damaging impact of a flood has been reduced over the last three centuries in France, gradually moving from the need to control water mobility through the construction of protection measures (active protection) to a need to regulate the residential mobility of people based on the likelihood of flooding in a territory (passive protection). This evolution has transformed protection against floods from an individual preoccupation to a collective issue of public security. Linking the local and national levels, this marked the multiscale character of floods and the institutional response – floods affect a broad range, from streams to rivers, and may be required from the commune to the State level. These two "hierarchies" support the application of the principle of subsidiarity of public action with regard to flooding. Broadly, this chapter summarizes the stream of technical and societal ideas that guide the French institutional response to flooding.

Chapter 6: The Difficulties in Local Implementation of Risk Prevention Policies

Flood risk affects half the communes in France and a quarter of the residential population. Faced with catastrophic situations, many risk prevention legal measures have promulgated since the 20th Century. In short, the development of urbanism in highly exposed flood-prone areas is no longer possible. However, urbanization continues to experience rapid growth. Often, natural risk prevention policies have been applied slowly and /or too late in exposed territories, often triggered by a catastrophic event. This chapter seeks to contextualize the co-existence of regulation and urbanization. Through this lens, the Gard department (in south-east France) is exemplary: recurrent floods, numerous decrees for natural catastrophes, exposed populations, etc. The example of Gard allows us to address the influence of major events on recent urbanization and on the evolution of legislation. It also allows us to understand how legal constraints have been taken into account in a differentiated way across territories.

Chapter 7: The Influence of Residential Migrations on the Evolution of Populations Exposed to a Flood Risk

The vulnerability of populations facing a flood risk is seen here over a long timeframe (several decades). After identifying the different factors that have triggered the residential migrations within the Gard department, the analysis turns to the evolution of the exposed population in 33 communes since 1968. It is shown that, except for the communes in the Cévennes that were severely affected by a decline in mining and traditional activities, all the communes saw an increase in the exposed population. However, important differences were noted between communes with comparable migration profiles. Three types can be identified: communes that quickly controlled the evolution of the exposed population without a need for regulations, those that, early on, were affected by the application of regulations aiming to limit settlement in flood-prone areas, and those whose decrease in exposed populations is recent and seems tightly linked to the establishment of flood risk prevention plans.

Chapter 8: Exposure to Flash Floods: The Conflict Between Human Mobility and Water Mobility

Human exposure to flash floods is seen in this chapter as a conflict between human mobility and water mobility. During a crisis, the difficult transition from daily activities to protection actions determines the response time of exposed populations. Between two crises, alternation between attitudes of resistance and acceptance appears in the temporal evolution of construction in flood-prone areas and in legislation. The period following a crisis brings multiscale interactions together, mixing the short and the long term.

This multiscalar observation puts forward three main operational recommendations. The first is to better understand the conjunctions that lead to the extensive institutional and policy repercussions following extreme events. The second suggests addressing the local occurrence of extremes that, although deemed "ordinary", have an impact the adaptation of mobilities. Finally, the third recommendation encourages a focus on the alternation between active and passive protection at different physical and social scales.

1

Mobility Exposure Scales of Analysis in the Face of Flash Floods

1.1. Introduction

As indicated in the Preface, the question of scales is essential to the articulation between water mobility and the mobility of individuals. The hypothesis of this book is that this notion contributes to understanding the emergence of critical situations with a hydrometeorological origin. Thus, we have explored several bibliographical references, which have furnished our reflection on the role of scales in the development process of the hydrometeorological event[1]. First, we worked with articles from Holling [HOL 01] and Cash *et al.* [CAS 06], resolutely oriented toward the interactions between environment and society. Then, we sought to deepen the way in which this notion is involved in the analysis of purely social phenomena, more specifically in sociology, by focusing on Grossetti's article [GRO 07].

The aim of this chapter is to provide a synthesis of our work and to explore the notion of scale, so as to better understand the links between physical and social dynamics when flash floods occur. First, we will show how scales can help us to analyze the dynamics at stake (section 1.2). Then, we will see which scale categories we need to mobilize in order to address this problem (section 1.3). Sections 1.4 and 1.5 will highlight the contributions of a multiscale approach for analyzing the interactions between man and the environment, on the one hand, and the scale approach toward social phenomena, on the other hand. Finally, we will propose and

---

Chapter written by Céline LUTOFF, Jean-Dominique CREUTIN, Isabelle RUIN and Sylvie DUVILLARD.

1 These bibliographic references helped us to build a common scientific culture within the interdisciplinary project *MobiClimEx*, on the one hand, and the AMETIhST seminar, on the other hand (see Preface).

discuss a synthetic scale, the scale of rhythmicity (or timeliness), helpful for understanding the combination of physical and social dynamics (section 1.6).

1.2. Why the question about scales?

Flash flood phenomena are fundamentally multiscalar. Several spatial and temporal scales intertwine when these phenomena unfold. Within a few minutes, small watersheds (a few square kilometers) are the first to react to intense rainfall. Some watersheds, dry and usually invisible, can be brutally burdened by strictly localized storm rains and sometimes cause violent overflows. Witnesses to the brutal reaction of the streams speak of walls of water or waves in order to express what they experienced. Larger streams usually have the capacity to absorb this type of narrowly localized rainfall. However, they are highly reactive to other weather phenomena on larger scales. Swollen by the contribution of this extreme rain to which the waters of smaller watersheds are added, their overflow takes place later, in a several-hour time span[2].

This overflow asynchronicity for a single and unique event was clearly shown in the work of Creutin *et al.* [CRE 13]. In particular, this article pointed out the way in which certain spaces, located at the junction of different size watersheds, can be confronted by several successive outflows within intervals of a few hours throughout the same hydrometeorological event. This is exactly what happened in Remoulins in the Gard department on September 8 and 9, 2002. During this historic flood, the east of the city was first flooded by the Valliguière, a tributary of the river Gardon, responding to an initial convection cell. Twelve hours later, the Gardon came to flood the city once again, in its western part this time, as a reaction to a set of wider thunderstorms, at a regional level. This double flood surprised both the inhabitants and the authorities, who were far from thinking that such a concomitance was even possible. But this conclusion was reached without taking the phenomenon of asynchronicity into consideration, as well as the successive reaction of different size watersheds to specific meteorological objects. Therefore, understanding the phenomena of flash floods and their dynamics involves dealing with the question of scales, particularly concerning the interaction between spatial and temporal scales.

As difficult as they are to predict, and in virtue of the reasons mentioned above, these brutal phenomena surprise people in their daily activities, leaving them very little time to react and organize their protection. Thus, understanding the exposure

---

2 In fact, all floods are multiscalar, but for slower floods, the reactive differences between watersheds are less perceptible. In those cases, the time reaction overlaps for some days or even weeks, whereas in the case of flash floods, we find watersheds, which react in a matter of minutes, while others respond in a few hours.

conditions of these populations also requires considering their mobility in time and space. During the event, how many people are exposed at home, at their place of work, performing related activities (hobbies, administrative procedures, shopping, etc.), and how many of them are traveling between these different places? Considering the exposure of these populations presupposes being able to reconstitute their mobility, taking into account the variations of spatiotemporal practices in time (between the days of the week, at the weekends, holiday periods), and also depending on the social characteristics of the individuals. Indeed, according to Massot and Orfeuil [MAS 05], mobility practices differ according to four main parameters: (1) the social parameter, which refers to the family structure of the individuals, their specific social interactions and constraints; (2) the spatial parameter, corresponding to places of residence and activities that induce specific time lengths and commuting time; (3) the cost parameter, implying both a temporal and a financial cost; (4) and finally, the means of transport (motorized or non-motorized mobility, use of public transport, etc.).

These mobility practices are part of the evolving dynamics of urban, suburban and rural spaces, which demand a multiscalar assessment in order to understand all the dimensions. In fact,

> "The process of urban sprawl and the resulting issues involve a multitude of scales which, in general, are not taken into account simultaneously: from the plot of land and its surrounding environment, which is the concern of the inhabitant [...] to the broad urban area the structure of which influences, for example, accessibility within the urban area or accessibility to certain amenities" [TAN 10].

So, understanding daily mobility requires defining residential choices in parallel, as well as their place in broader territorial dynamics.

Residential choices depend on a number of factors, which, at the same time, depend on very different scales. Previous studies carried out on around 20 cities in different Northern and Southern countries reveal some recurring factors. Among these, we particularly pay attention to the distance to work places or the schooling possibilities for each member of the household, the accessibility to various modes of transport developed at local level, the proximity to certain places of activity, such as shopping centers, recreation sites, etc. [BON 00]. For certain households, workplaces or study sites, to discuss only two, can involve very variable scales: from the local schooling options (the district, the residence municipality) to the distance to the central business district, and sometimes even more, as in those cases where the couple may work in two different cities.

For a given space, these local mobilities represent an important part of the daily traffic using various modes of public and private transport. In addition, there are trips recorded on a much larger scale and for which the location in question is only an obligatory throughout for example a major interchange. These complementary mobilities concern the transport of goods, certain professions, and certain transfers, which may take place on a regional, national or international scale. These other types of mobility add even more scale ranges.

Thus, understanding the impact of flash floods on mobility presupposes a crossing between the dynamics of the phenomenon, on the one hand, and daily and residential mobility, on the other hand. This requires a special consideration of the multiple scales involved. As we have just seen, spatial and temporal scales are essential because of the very nature of the phenomena observed. However, other scales can be mobilized to better understand how hydrometeorological and social dynamics interact during a particular event. We now aim to identify which scales can be mobilized so as to give a relevant account of what is at stake in this interaction.

1.3. Which scales should be mobilized?

As we have just seen, the consideration of spatial and temporal scales becomes essential in addressing the interactions between daily or residential mobility processes, and hydrometeorological dynamics. Therefore, a hydrometeorological event is the result of the conjunction in time and in space of two dynamics: the physical and the social (according to the definition that Walters [WAL 12] provides for the term event[3]). Nevertheless, in order to understand what is at stake in these events, how they are influenced by the past and what they are capable of provoking in the future, other reference scales must be examined. Thus, there are individual and collective decisions in relation to this event, mobilizing specific levels of organization or of social interactions that should be taken into account.

Testimonies collected during post-event investigations (PEIs) show, for example, that the decision to travel or not is not made under the same conditions, depending on whether the person is a resident or passing through, alone or accompanied, particularly by children [RUI 14]. Experience in similar events and family interactions seem to be a major component for decision making in critical situations. Similarly, the reaction of a community (a district or a municipality, for example) follows a regulatory framework more or less dependably, taking into account the interventional capabilities of other institutional levels. Thus, the French *Plan Communal de Sauvegarde* (Municipal Protection Plan, PCS) defined at a municipal level, can be triggered locally,

3 "An event is defined simply as something that happens somewhere during a particular interval of time" [WAL 12, p. 135].

with the aim of evacuting certain areas. However, if the event exceeds the boundaries of the municipality, the PCS is articulated with the implementation of the Organisation de la Réponse de Sécurité Civile (ORSEC) plan, at a departmental or regional level. Implemented actions are coordinated between these two levels of intervention.

After a crisis, when we can now consider the longer term consequences of hydrometeorological events, particularly about the organizational choices of territories, other reference scales must be considered. Thus, residential choices are based on dynamics, which are not exclusively spatial or temporal. Thomas and Pattaroni state the hypothesis that:

> "the residential-location trade-offs that middle-class families make take into consideration not only a balance in terms of price and housing size, but also the functional, perceptive and social factors that determine the quality of life at a given place. Thus, there exists a vital link between residential choices, family lifestyles and the characteristics of the built environment" [THO 12, p. 112].

Here, the term environment is understood in the broader sense, expressing both functional (transport, accessibility, etc.), perceptive (noise, sight, calm, etc.) and social aspects (neighborhood quality and interrelations). The observation of Switzerland's middle classes made it possible to identify specific residential profiles (this was a way of assigning value to goods), depending on the types of environments taken into consideration. The authors showed that residential areas, their tenure and prices, vary positively when they are near certain environmental factors. However, other factors can negatively affect these spaces and their value. Faburel and Gueymard suggested measuring this possible negative impact of the environment on residential choices [FAB 08]. For this, they performed studies on environmental inequalities. These studies were implemented as a part of the *Schéma Directeur d'Ile de France.* They suggested a classification of the environmental factors identified according to the level of influence they have on the setting up of certain categories of residents, thus identifying so-called environmental inequalities. Their ranking of negative factors places industrial risk at the top of their list of problematic environmental characteristics and, in decreasing order, air traffic-related noise pollution, road or rail traffic, the risk of flooding and car pollution. We see here that the risk of floods constitutes one of the possible negative aspects in residential choices, but that it is placed in the penultimate position in the ranking of negative factors. The advantages of being close to a watercourse, especially from a landscape point of view, can potentially explain this classification in cases where floods are rare. One wonders, however, if the study would have yielded the same results just after the floods of the Seine and its tributaries in the spring of 2016. While this factor remains relative compared to the other components taken into consideration in this research, a deeper analysis nonetheless shows that flooding is clearly taken into account in what concerns socio-urban profile differentiation. In fact, the distribution of these profiles in flood-prone areas (FPA) shows that 54.5% of the inhabitants come from modest profiles, compared to 19.5% for average profiles and 26% for the better-off, whereas the overall profiles distribution in Ile de France is, respectively, 33.15%, 34.69% and 32.15% for the period studied.

So, it appears that the scales of environmental approval, or, on the contrary, disapproval, have an important role in residential choice and, consequently, in the assessment of residence places. Therefore, evaluating the impact of a flood on the evolution of the notion of value implies placing this factor in relation to other reference scales involved in the process of residential mobility (especially accessibility levels, landscape and social interactions). Going further in identifying the reference scales to be taken into consideration in the different modes of interaction between man and the environment, we will turn to the synthesis proposed by Cash *et al.* [CAS 06].

1.4. Multiscalar approach to the interactions between man and environment

The article by Cash *et al.* [CAS 06] offers the hypothesis that the failure of public policies related to environmental issues might be linked to a lack of consideration of different scales and interscale dynamics for understanding the various forms adopted by the interactions between man and the environment (pollution, vulnerability to floods or drought, epidemics, etc.). Inspired by Gibson *et al.* [GIB 00], the notion of scale is defined as "the spatial, temporal, quantitative, or analytical dimensions used for measuring and studying any phenomenon". On the other hand, the notion of level concerns "the units of analysis that are located at different positions on a scale". Alongside spatial and temporal scales, the best known and the most thoroughly studied, Cash *et al.* [CAS 06] also introduce a *jurisdictional scale*, close to the spatial scale, but which makes it possible to clearly define the limits of each political unit (a commune, a department or a county, a region, etc.). It also differs from the institutional scale, which is related to law hierarchy and collective rules. There are also other scales proposed by the authors in order to better understand the interactions between man and the environment, such as the management scale, which makes a distinction between the following levels: the task, the plan or the strategy. The *network scale* refers to certain interactions, which do not directly relate to spatial or jurisdictional (political) considerations, such as markets, community networks (religious ones, for example), or professional or associative networks. Finally, Cash *et al.* [CAS 06] suggest a *knowledge scale* that makes it possible to differentiate knowledge systems, which are rarely connected, such as scientific knowledge, on the one hand, and experience-based knowledge, on the other hand. Science seeks to provide knowledge on a global scale, sometimes difficult to transpose to the local scale. Local knowledge is essential for understanding certain human–environment interactions and for solving certain specific problems.

Figure 1.1 suggests placing flash flood phenomena, as well as daily and residential mobilities, on these different scales. In terms of space and time, we observe a concomitance between flood phenomena and daily mobilities: these dynamics both concern a spatial level that is locally restricted and a temporal extent, regarding a daily schedule. However, daily mobility may go beyond this spatiotemporal frame and involve higher levels. For example, this is the case of individuals living in one city and working in another one (Nîmes and Paris, for instance). Nevertheless, in this case, mobility is no longer daily and mobility rhythms are adapted to distance constraints and trip length. Residential dynamics are clearly differentiated on the spatial and temporal scales, which can extend to far larger levels.

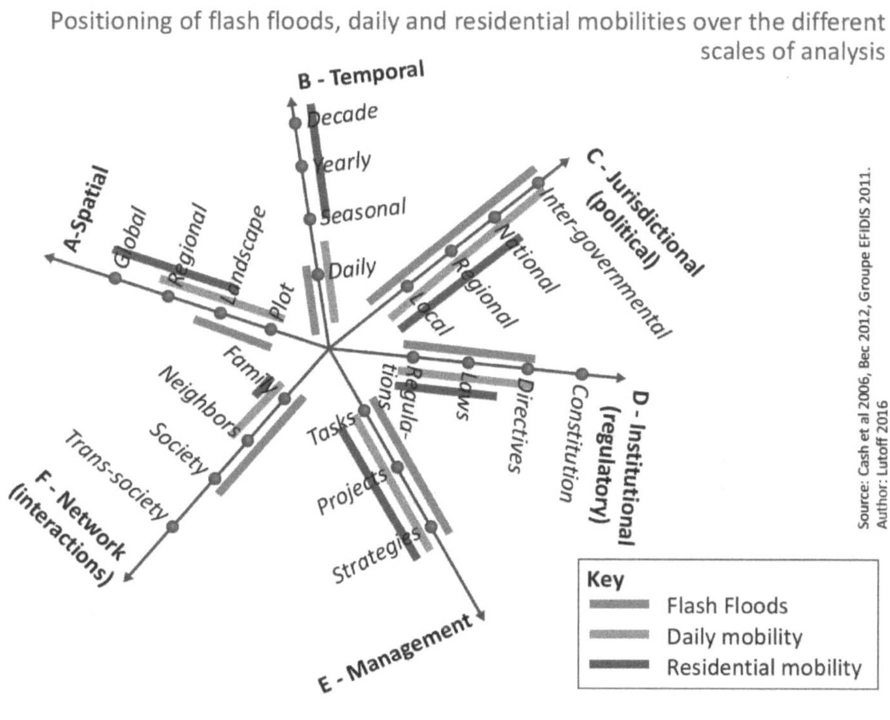

Figure 1.1. *Positioning of relevant levels of analysis regarding flash floods as well as daily and residential mobilities over the different scales suggested by Cash et al. [CAS 06] for analyzing the interactions between man and environment. For a color version of this figure, see www.iste.co.uk/lutoff/hydrometeo1.zip*

Jurisdictional (political decision making) and institutional scales (marking the hierarchy of norms) involved in the three observed phenomena are quite similar. Certain political decisions and regulatory standards regarding flash flood management and daily mobility derive from European directives, which can be transposed to French law, in compliance with the constitution [CEP 12, LEG 12]. As far as we are concerned, political decisions regarding flood risks fall under the European Union Flood Directive promulgated in 2007, which imposes a number of measures on Member States so as to improve risk assessment. The transposition of this directive into French law was made through the law of 12 July, 2010 (known as Grenelle 2), comprising six components, including one on risk prevention. These regulatory mechanisms made it possible to define the National Strategy for Flood Risk Management (*Stratégie Nationale de Gestion des Risques d'Inondation* [SNGRI]), promulgated in October 2014, and to identify Flood Risk Areas (*Territoires à Risques d'Inondation* [TRI]) in which Management Plans (*Plans de Gestion des Risques Inondation* [PGRI]) must be defined at the level of hydrographic districts (or large watersheds, of which there are eight in metropolitan France).

Concerning daily mobility and trips, the measures enforced fall under European directives of the 1970s aimed at the reduction of greenhouse gases [HUG 06]. Their implementation in France also falls under laws Grenelle I and II, of 2009 and 2010. Grenelle I focused on the development of Urban Mobility Plans in French agglomerations. In 2010, Grenelle II law reinforced the development of public transport and the definition of Priority Action Zones for Air (*Zones d'Actions Prioritaires pour l'Air* [ZAPA]) "banning the circulation of vehicles emitting the highest amount of particles and nitrogen oxides" in certain experimental territories [ADE 12].

Regarding residential mobility, however, for the time being Europe does not seem to be directly involved in the implementation of regulatory measures. The latter are mainly based on decisions made at the national or local level, particularly in order to contribute to the development of affordable housing for all. This was the case, for example, of the MOLLE law (*loi de Mobilisation pour le Logement et la Lutte contre l'Exclusion*), favoring housing facilities against exclusion, dating from 25 March, 2009. The aim of this law was to "make social housing more fluid" (EFIDIS Group, 2011).

Finally, the network (or interaction) scale highlights the differentiated positions of the three processes we are studying. In fact, the interactions involved in a hydrometeorological event extend from the family unit (an essential unit in this type of situation) to the whole of society, when the event is strong enough to trigger the modification of practices far beyond the impacted area. This was the case, for example, for the Xynthia storm, which entailed consequences in the organization of territories beyond the crisis itself and the affected areas, and provoked the taking into consideration of risks by all of French society[4]. On this scale, daily mobility mainly concerns the family unit and close relatives: the neighborhood, for example,

4 This topic will be developed in Chapter 5.

through the development of carpooling. Finally, residential mobility generally involves the family unit exclusively, except when it concerns the transfer of entire population groups (currently, migrants, for instance). Nevertheless, we will later discuss the topic of migration, which induces a type of residential mobility, but differs in that it involves very specific causes (mainly political ones).

While the network scale makes it possible to address some of the interactions staged by a flash flood event, it nonetheless remains limited. For example, how should we position the interactions that emerge at the moment of the event? PEIs show examples of very specific interactions at the time and place of the flood. These interrelations existed only for this particular moment, which cannot be attributed to the family unit or to the relatives. Therefore, the network scale does enable taking them into account while they may be crucial for the safety of certain people at the time of the crisis (particularly by encouraging them to leave certain places in order to seek refuge elsewhere).

After this exploration of the multiscalar approach, we can say that the scales suggested by Cash *et al.* are crucial understanding the spatial and temporal dimensions of the interactions between flash floods and mobility. They also include legal, institutional and management scales, which are interesting for understanding the social administration of these interactions. However, when we focus on the social interactions generated by floods at different scales of time and space, the sole use of the network scale appears a limitation. This is the reason why we will turn to the proposition that Grossetti [GRO 07] advanced for scaling social phenomena, which will undoubtedly complete our approach.

1.5. Social phenomena scales

This section evaluates whether the scales proposed by Grossetti [GRO 07] are applicable to the social phenomena that we observe, and thus completes the multiscalar approach suggested by Cash *et al.* [CAS 06]. In his article, Grossetti outlined a vision of social phenomena around three large scales: mass, duration and generality.

The *mass scale* refers to the levels of interaction involved in the observed phenomena, ranging from an individual to a massive level, via the organizations and particular action networks. In a context of flash floods, daily and residential mobilities are staged at these different levels. As we have previously indicated, individuals can be confronted with extreme natural phenomena in their daily life, particularly during their daily commute. This means that they have to manage the event in a context of limited but essential interaction. In fact, post-event investigations (PEIs)[5] [RUI 14] have shown that these situations are quite common. Some witnesses have reported that they mainly interact with relatives that they contact by telephone. Others have referred to what we denote "misfortunate groups", constituted at the time of the crisis and in random situations. In several cases, the constitution of these spontaneous and ephemeral groups enable an effective protection of the persons concerned, some of who could have been killed had they remained isolated.

The *duration scale* can be compared to the time scale by Cash *et al.*, though adapted to social phenomena. It is divided into three levels: short time (immediacy), biographical time (less than a human life's duration) and historical time (which can aggregate several human lives, that is, several generations). Applied to our problem, it makes it possible to distinguish different social dynamics. At the first level of the scale, we find individual reaction or misfortunate groups immediately after the event. In biographical time, we distinguish residential trajectories, which may be more or less influenced by the occurrence of floods. Traumatized by an event, some individuals may choose to change their place of residence so as not to be confronted with this type of event again.

However, as we will see in Chapter 2, these cases are very limited. Moreover, as we mentioned above, the evolution of residential areas in flood-prone areas (FPAs) and the associated real estate market can be influenced at this duration level, due to flood recurrence. One of the aims of the ANR *MobiClimEx* project is to understand precisely this influence better and evaluate it over time. Finally, on a historical scale, it is the memory of past events, which contributes to a risk culture and to the management of its evolution in the long term. The creation of SCHAPI[6] in 2003, for example, took place at this level. Arising from a better understanding of the phenomena and the necessary conditions for improving their forecast, this organization aims to provide technical support to regional forecasting services, as well as to ensure their coordination. For the general audience, it complements *Météo France's* weather forecast by proposing hydrological forecasts for main waterstream areas.

---

5 Post-event investigations (PEIs) are surveys and field work carried out after an event in order to collect useful data for analyzing the occurrence. Sociohydrometeorological PEIs marry the competencies of hydrometeorologists and social sciences experts to jointly collect invaluable data associated with their different analysis dimensions (see [RUI 14]).

6 *Service central d'hydrométéorologie et d'appui à la prévision des inondations* (Flood Supporting Forecast Hydrometeorological Centralized Service) is a French nationally positioned unit responding to the French Ecology, Sustainable Development and Energy Ministry (*Ministère de l'Écologie, du Développement durable et de l'Énergie* [MEDDE]). The SCHAPI establishes and publishes ongoing information alerting about possible floods at a national level in coordination with the flood forecasting services.

The third scale, the *generality scale*, concerns specificity or generality levels of the "action domains" involved in the phenomena observed. It makes it possible to measure the number of contexts taken into account. The first level, a highly specialized one, is related to a single social context. For example, this is the case when we are interested in the interactions generated by the emergence of misfortunate groups or during the implementation of the communal relief plan within a municipality. At this level, interrelations between the actors are limited to the people gathered at the time of the event or to what is foreseen by the plan. The second level concerns the interaction of a limited number of contexts. Such is the case when we consider the evolution of the regulations related to risk assessment associated with urbanization, for example, or even when we are interested in the manner in which the real estate market (and consequently, residential mobility) becomes more or less influenced by hydrometeorological events and risk prevention regulations. The third level concerns what is generally classed under the macroscopic scale, involving "every imaginal context" [GRO 07]. In our case, this refers to the production and mobilization of the norm, be it in terms of risk management and planning, or for territorial development.

In order to explore this three-dimensional space of social phenomena, Grossetti [GRO 07] suggests employing double input iterative tables, neutralizing one of the three dimensions each time, so as to focus on the other two. We carried out the exercise by repositioning an example of a social process involved in the interactions between mobilities and flash floods. We chose to work on adaptation behaviors regarding the flood: which is the characteristic behavior toward the event? As planning practices evolve, what is the place of the appraisal of these occurrences?

Table 1.1 depicts these processes according to mass and duration scales: in black, individual and collective adaptation at the time of the flood; and in gray, flood management practices in FPA. It also describes how these processes can be observed at these different scales. As we mentioned earlier, individual and collective behavior in the face of floods are related to mass levels 1 and 2 (interactions and organization), which can manifest in the whole duration of range levels, from the very short time of the crisis to historical time, regarding the evolution of this behavior in the long run. Management practices in FPA are deployed over a different space, and are instead related to higher levels of mass and length: organization and mass develop throughout biographical and historical time.

Table 1.2 focuses on collective adaptation practices, through crisis management or the management of FPAs, with particular emphasis on the development of regulations for acknowledging flood risk. As we have seen, time is a major element of analysis; we have also chosen to explore the impact of the generality scale on these practices. By means of this analysis, we would like to see how the evolution of collective management and planning practices of FPAs have repercussions on the evolution of the real estate market, and thus, in turn, on residential mobility.

Duration Mass	(1) Short term, very short term, Upsurges	(2) Biographical time (< human life) Aggregates	(3) Historical time (> human life) Sedimentation
(1) Interaction	Individual behavior in critical situations: (SHM PEIs), *online research, videos, photographs, testimonies*	Individual behavior models in critical situations: *series of individual SHM PEIs.* Life trajectories in flood-prone areas: *ethnographic analysis*	Evolution of risk-related behavior models: *photographs, engravings, texts, correspondence regarding flood events*
(2) Organization, action system, primary network	Organizational behavior in critical situations: *SHM PEIs in communities. Short-term monographs.*	Organizational behavior in critical situations: *series of collective SHM PEIs, series of online investigations on crises.* Risk acknowledgment in urban agglomerations: *comparison of urbanism documents on two different dates.*	Mobility adaptation in the face of floods: *Evolution of recorded floods, historical analysis of the deployment of human settlements, urban layout analysis.*
(3) Mass		Adaptation in the face of risk: *Yearly recorded statistical series (flow), tax statistical series (stocks), Catnat data*[7] *analysis.*	Evolution of relationships between society and floods: *Yearly recorded statistical series (flow), sedimentary archive study + analysis of historical archives.*

Table 1.1. *Process typology observed in the face of flash floods and observation methods according to mass and time scales: in black, individual and collective behavior in the face of floods; in gray, urban management practices. In italics, suggested plausible observation methods for understanding these processes. SHM PEIs: sociohydrometeorological post-event investigations*

7 The French Catnat is an insurance database concerning natural disasters.

Duration generality	(1) Short term, very short term	(2) Biographical time (< human life)	(3) Historical time) (> human life)
(1) Highly specialized (one context)	Decision/adaptation of activities and mobilities in the face of a flood: *service and emergency action; ORSEC Plan triggering; alerts; evacuation. Self-organization of misfortunate groups* *Post-flood crash sales*	Post-flood local risk management: *tool mobilization for limiting future impact (urbanism documents in flood-prone areas (FPAs), risk management tools)* Evolution of the real estate market in the post-flood territory Post-flood migratory residential dynamics Individual residential trajectory	Evolution of flood risk calculation over a territory in the long run: *evolution of layout risk calculation and mobilization of regulatory tools so as to locally manage risk.*
(2) Specialized (limited number of contexts)		Joint evolution of urbanism regulations and risk acknowledgment: *PPRI*[8] *implementation, PLU*[9] *revision* Risk assessment in the evolution of the real estate market Migratory residential factors: *natural risk assessment regarding the evolution of local employment…*	Evolution of risk calculation in territory management Development of a national risk prevention policy
(3) Generic		Regulation production and mobilization Credit levels and construction costs	Great social evolution: access to property, female employment, relation to risk, "South" effect and impact on the real estate market

Table 1.2. *Typology of collective adaptation practices to flash floods according to mass and generality scales: in gray, management practices in flood-prone areas (namely by means of regulation evolution); in black, the evolution of the real estate market (major factor for residential mobility evolution). In italics, possible observation modes for understanding these processes*

8 PPRI: *Plan de Prévention des Risques d'Inondations* (Plan to Prevent Natural Hazards of Flooding).

9 PLU: local urban planning in France.

The exercise shows that the processes we are studying are distributed in a relatively uniform manner along the length scale, representing a strong axis of the interaction between physical phenomena and social dynamics. Once again, resorting to the generality scale makes it possible to identify specific social interactions that emerge with the event, and to differentiate them from more elaborate processes (implementation of crisis management plans, for example). Applied to medium and long-term processes, the use of the generality scale makes it possible to position emerging tools and processes after a flood in a given territory. Later, it will be possible to generalize these tools and processes in the form of regulations applicable to all flood zones. This analysis method applied to the real estate market will also make it possible to introduce questions about the pace of evolution in property values, and to analyze certain major trends in FPAs.

The social phenomena scales suggested by Grossetti [GRO 07] make it possible to enrich our analysis. Here, we have assigned a prominent place to the time (or duration) scale in the application of this approach to the interactions between mobility and flash floods. However, the spatial scale, equally central to the case that interests us, has no place in Grossetti's proposal [GRO 07]. It is the aim of acknowledging this need that we now suggest representing the different spatiotemporal levels involved in flood events on the same scale.

1.6. Rhythmicity scales: spatiotemporal summary of the observed dynamics

Here, we seek to represent the spatial and temporal dimensions of hydrometeorological phenomena, which impose themselves on social dynamics as unavoidable constraints, on a single scale.

Coined as "timeliness", a first draft of this specific scale was proposed in [LUT 16] and renamed here as rhythmicity scale. Its definition is based on three initial hypotheses: (1) during the crisis, the flood's peak is considered as the moment in which waters reach their highest level. It marks the end of the anticipation period and the peak of danger; (2) it is essential to correctly estimate the anticipation time remaining before the peak of danger in order to understand how much time is left for getting prepared against the flood. We can easily understand that taking protective measures is not done under the same conditions, depending on whether protective action is conducted before or after the peak of danger; (3) chronological time must be underestimated according to the water's rising rate specific to the watershed in which the action is taking place. The level of anticipation of an action does not have absolute significance, but must be underestimated according to danger's irruption levels, for instance, the rising rate of the waters according to the considered watershed.

The time a river takes to reach its flood peak depends on its load rising capacity. For evenly distributed rain at the spatial level, the time will mainly depend on the size of the watershed and average river speed. Zoccatelli *et al.* [ZOC 10] showed that the rapidity of rise in water depends on the spatial organization of the precipitation field and the morphology of the hydrographic network. In their analysis of flash floods, Marchi *et al.* [MAR 10] indicated that the time between maximum rainfall and the peak flow represents the square root of the watershed's size up to 350 km^2, and then increases almost linearly in larger areas. In addition, the delay between the start of a flood-generating precipitation and the peak of floods also depends on the size of the watershed, which expresses a kind of resonance between atmospheric and hydrological scales. According to Creutin *et al.* [CRE 13], the response time of the watershed in relation to the onset of the rainfall can be calculated by applying a 2.6 factor to the starting-point equations proposed by Marchi *et al.* [MAR 10]:

$$T_L = 0.08 \times A^{0.55} \text{ for watersheds} \leq 350 \text{ km}^2 \quad [1.1]$$

$$T_L = 0.0032 \times A^{1.10} \text{ for watersheds} > 350 \text{ km}^2 \quad [1.2]$$

Large watersheds are characterized by long response times that make flood management easier. For example, in January 1910 in Paris, the Seine (44,000 km^2) reached its flood peak in seven days, after 13 days of precipitation [MAR 97]. During the more modest flood that hit the Parisian region in early June 2016, the rise of the Seine was achieved in eight days, which made it possible to evacuate 20,000 people and several museums (Le Monde, June 3, 2016). Nevertheless, during a flash flood rivers respond much more quickly. In June 2010, in the Var, the Nartuby reached its peak in Draguignan (184 km^2) in just two hours, after 3–4 hours of intense rain [RUI 14]. As mentioned earlier, the combination of highly localized storm cells and the rapid concentration of water in small watersheds create asynchronous responses characteristic of a multitude of local flood scenes randomly generated over fairly large areas, covered with rainfall accumulations of more than 200 mm/h [CRE 13].

Thus, the major difference between a river flood and a flash flood essentially lies in the number of specific flood scenes and their particular tempo. A flood of the Seine in Paris generally obeys the unique rhythm imposed by the main river, which determines water levels. In flash flood situations, we are confronted with a myriad of flood scenes, distributed in space, each of the flows being sufficient to threaten human lives. In 100 m, dangerous conditions can vary fundamentally. It is therefore essential to be able to define anticipation conditions for each of these scenes[10].

Rather than using traditional spatiotemporal benchmarks to assess the actions undertaken in the face of a flood, we should evaluate the anticipation levels of individual or collective actions for each flood scene. The anticipation level of the actions is evaluated on this new scale, the rhythmicity scale, in which there are four stages: (1) actions undertaken in the absence of any visual precursor, only on the basis of climatological precursors; (2) actions with regard to meteorological precursors (heavy rainfall); (3) actions undertaken facing hydrological precursors (first overflows), but before reaching the flood's peak; (4) actions implemented after the flood's peak are considered not anticipated, and belated. These different levels summarize both the dynamics of the watershed and the unfolding time of flood over a specific scene. Levels vary from scene to scene and from event to event. They define the tempo that the river imposes on anticipatory actions, as if it were an environmental stress factor.

Defining the precursor span that characterizes each level of the scale, we rely on Torny and Chateauraynaud, who have connected the signs that can launch preventive actions [CHA 99]. Climate precursors are useful when no concrete sign of the crisis is still visible during the build-up phase well before the event. Different climatological tools, expert or layman, can surely be summoned. Herein, the term climatology is used for referring to climatological science, which provides the basis for meteorological and hydrological forecasts. However, it also refers to a climatology carried out "in an amateur manner" (in the noble sense of the term), with a careful observation and in-depth knowledge about the everyday environment, which certainly favors the development of specific forecasting and prevention capabilities. Meteorological precursors concern intense rainfall. Given the complexity of the transfer between rain and water flow in the rivers, these are difficult to interpret. However, PEIs show that an abnormal intensity of rain can be the triggering point for some preventive actions, especially when extreme hydrometeorological situations occur in a recurring manner. The actors and observers of the territory are then able to identify rain, which can give rise to flash floods. Hydrologic precursors refer to rising water levels in streams and rivers, which may, again, initiate specific actions before reaching the peak of danger.

In order to render a comparison between anticipation levels or characteristic reactions to the different flood scenes, the anticipation scale is expressed in relative values. The main unit we use for ranking the position of undertaken actions on this scale is the typical response time of each river near which the action took place. This average

10 See Chapter 3 and Borga *et al.* [BOR 11].

time response can be defined as the difference between the rain's peak and the flood's peak [CRE 13].This unit is therefore the specific amount of time required for a watercourse to respond to a peak of rain (to rise its load) and to generate a flood. Thus, it is the available amount of time from the moment when the rain becomes most intense, for reacting and anticipating a potential flood. We suggest ranking the undertaken actions according to a relative value, corresponding to a part of the physical amount of time available at the precise setting of the action. This relative scale makes it possible to compare different anticipation capacities in various flood scenes measured with the aid of a unique scale of value. So, the anticipation level of an action is calculated as follows:

$$AT - FPT = AntT_{absolute}, \qquad [1.3]$$

where:

– AT = action time,

– FPT = flood peak time,

– $AntT_{absolute}$ = absolute value for anticipation time.

The relative anticipation time is calculated as follows:

$$\frac{AntT_{absolute}}{R_T} = AntT_{relative}, \qquad [1.4]$$

where:

– $AntT_{absolute}$ = absolute value for anticipation time,

– R_T = watershed response time,

– $AntT_{relative}$ = relative anticipation time.

As we mentioned earlier, there is a relationship between the size of the watershed and its response time. In order to estimate the relative response time of the watershed in relation to the beginning of the rain (and no longer in relation to the precipitation peak), we applied a factor of 2.6 to the watershed response time previously calculated according to the recommendations of Creutin *et al.* [CRE 13].

Figure 1.2 presents the theoretical situations that we may encounter and the discretization mode used for associating a precursor category with every possible action.

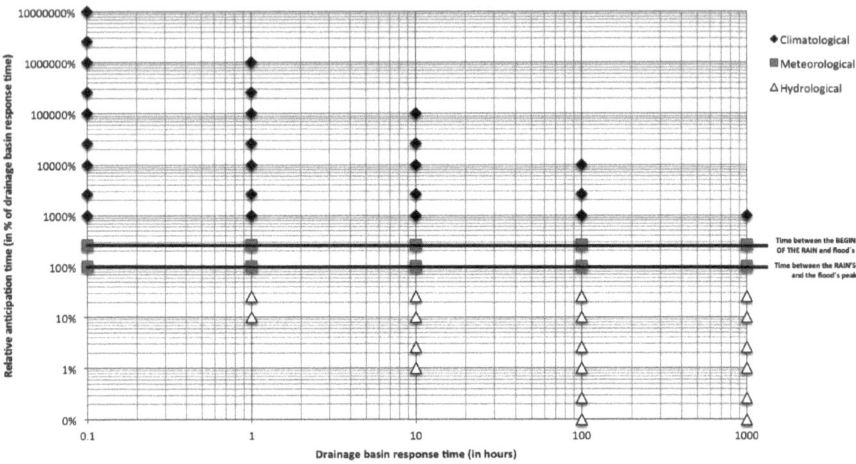

Figure 1.2. *Precursor type definition linked to a theoretical action depending on the response time of watersheds in which the action can be taken and the calculated available anticipation time (according to [LUT 16])*

A 100% relative anticipation time means that the entire response time of the watershed on which we are located has elapsed when the action is taken. If the anticipation time is shorter (closer to 0 and to the flood's peak), then it is likely that the action will be undertaken at the moment when hydrological phenomena are observable (first overflows). Action carried out during the time frame of relative anticipation situated between 100% and 260% is achieved at the moment when the waters are already rising (either from the beginning of the rain or at the moment when rain has reached its maximum intensity). Therefore, the essential precursor in this case is the rain. The actions undertaken during the build-up of this relative time frame can be considered as highly anticipated, since visible indicators are not present and it is necessary to resort to climatological precursors (scientific or not), in order to envision a possible flood.

The rhythmicity scale then makes it possible to integrate the spatiotemporal constraint imposed by the hydrometeorological phenomenon on a unique axis, thus enabling the possibility of cross-referencing this piece of information with other dimensions featuring mobility and/or adaptation social processes.

1.7. Scales of mobility exposure to flash floods

At the end of this analysis, the notion of scale appears to be essential for understanding what is at stake during flash flood events, between the physical dynamics of the phenomenon and the social dynamics of mobility and exposure. The multiplicity of scales explored shows that (1) spatial and temporal scales are inherent to the studied phenomena and can be summarized by the notion of rhythmicity; (2) the scales suggested by Cash *et al.* are interesting for breaking down the different processes involved, but they have limitations; (3) the use of mass or generality scales favors a more refined approach to the social processes at work. According to these scales, it seems easier to differentiate between what is specific to the exposure to a particular event and what can be generalized to all FPAs.

The rhythmicity scale also suggests it is possible to take the specificity of the different flood scenes into account and to integrate the particular conditions imposed by the physical phenomenon within the analysis of the social responses. In previous works, the rhythmicity scale was applied to actions undertaken at the moment of the crisis to better understand whether the hydrometeorological context had an impact on the adaptive capacity of individuals or communities [LUT 16]. According to observations made in several cases in France and in Italy, it seems that while it is true that the hydrometeorological context plays a central role, it does not suffice to explain, by itself, the differences in the anticipation capacities of the populations confronted with the event. Thus, even in very small watersheds, characterized by very short response lapses, hazard anticipation is possible and regularly implemented. Canceling certain trips, protecting the property against the intensity of rain etc.; all these actions made it possible to reduce the impacts during the events observed. Contrary to this, decisions made in large watersheds and with a longer time for reaction does not necessarily mean that the actions taken will be anticipated more efficiently. A generalization of these conclusions is nonetheless impossible without further observation based on other cases. The systematic observation of hydrometeorological events and of the spatiotemporal conditions of social responses is essential for making progress in this direction. It also seems important to continue research efforts in exploring what may explain the differences between individual and collective response capacities, by exploring other, more social, dimensions of the processes involved.

We may ask whether the rhythmicity scale could be applied to the actions taken following a flood or due to the certain recurrence of intense hydrometeorological phenomena in a context of climate change. Can we observe a rhythm imposed by natural phenomena based on the adaptations performed over the longer term? In order to answer this question, it is necessary, on the one hand, to define a common temporal origin for comparing observed social processes. In the analysis of anticipation capacities, this common origin is represented by the flood's peak, which makes it possible to compare the behavior undertaken on very different flood scenes. When we seek to compare the social responses undertaken after the flood (such as adapting planning regulations or the evolution in the occupation of FPAs), we can define a reference event as the origin time. In the case of the Seine, this reference event has long been the flood of 1910, which also serves as a global reference at the national level, despite the specificity of its dynamics. In the south east of France, many events can be taken as reference points, according to the analysed territory: the flood of 1999 of the Aude, those of 1988 or 2002 of the Gard, that of 2010 of the Var, etc. Faced with this diversity, how can we define a reference event common to the different categories of actions? Answering this question implies questioning the notion of a high-impact phenomenon[11] and verifying which events were the starting points for specific actions in terms of adaptation to different levels of mass and general scales.

Having a common temporal origin for comparing adaptation actions can lead to the identification and characterization of events that have been significant in time and space, but also from the perspective of jurisdictional, managerial, mass or generality scales.

Adopting the rhythmicity scale for adaptation purposes requires evaluating the characteristic rhythm of action. In the case of anticipatory actions, this is imposed by the phenomenon itself. In the case of adaptive actions following a reference event, we may wonder whether this rhythm will naturally become social again, or whether the recurrence of floods will impose a new rhythm. So far, we do not have the necessary means for answering these questions. However, the scale approach adopted in this work regarding the exposure of daily and residential mobilities to flash floods will enable us to explore some useful leads.

1.8. Bibliography

[ADE 12] ADEME, L'écomobilité. Repenser nos déplacements, Report no. 7234, April 2012, Florent Bec/L-103 Communication, 2012.

[BON 00] BONVALE C., DUREAU F., *Les modes d'habiter: des choix sous contraintes. Métropoles en mouvement. Une comparaison internationale*, Anthropos, Paris, 2000.

[CAS 06] CASH D. W., ADGER W. N., BERKES F. *et al.*, "Scale and cross-scale dynamics: governance and information in a multilevel world", *Ecology and Society*, vol. 11, no. 2, 2006.

[CEP 12] CEPRI, nd, Politiques de prévention issues de la directive inondation et de la loi Grenelle II, 2012, available at http://www.cepri.net.

[CHA 99] CHATEAURAYNAUD F., TORNY D., *Les sombres précurseurs. Une sociologie pragmatique de l'alerte et du risque. Recherches d'histoires et de sciences sociales*, de l'École des hautes études en sciences sociales, Paris, 1999.

11 See Chapter 4.

[CRE 13] CREUTIN J.-D., BORGA M., GRUNTFEST E. *et al.*, "A space and time framework for analyzing human anticipation of flash floods", *Journal of Hydrology*, vol. 482, pp. 14–24, 2013.

[EFI 11] EFIDIS, Le parcours résidentiel, 2011, available at http://www.efidis.fr/.

[FAB 08] FABUREL G., GUEYMARD S., "Inégalités environnementales en région Île-de-France: le rôle structurant des facteurs négatifs de l'environnement et des choix politiques afférents", *Espace populations sociétés*, vol. 1, pp. 159–172, 2008.

[FRE 83] FRENCH J., ING R., VON ALLMEN S. *et al.*, "Mortality from flash floods: a review of National Weather Service reports, 1969–81", *Public Health Reports*, vol. 98, no. 6, pp. 584–588, 1983.

[GIB 00] GIBSON C.C., OSTROM E., AHN T.K., "The concept of scale and the human dimensions of global change: a survey", *Ecological Economics*, vol. 32, pp. 217–239, 2000.

[GRO 07] GROSSETTI M., "L'espace à trois dimensions des phénomènes sociaux", 2007, available at http://halshs.archives-ouvertes.fr/halshs-00477433.

[GUE 06] GUEYMARD S., "Facteurs environnementaux de proximité et choix résidentiels. Le rôle de l'ancrage communal, des représentations et des pratiques des espaces verts", *Développement durable et territoires*, dossier 7, 2006.

[HOL 01] HOLLING C.S., "Understanding the complexity of economic, ecological, and social systems", *Ecosystems,* vol. 5, pp. 390–405, 2001.

[HUG 06] HUGREL C., JOUMARD R., "Directives et facteurs agrégés d'émission des véhicules routiers en France de 1970 à 2025", 2006, available at http://hal.archives-ouvertes.fr/hal-00916989.

[JON 05] JONKMAN S.N., KELMAN I., "An analysis of the causes and circumstances of flood disaster deaths", *Disasters*, vol. 29, no. 1, pp. 75–97, 2005.

[LEG 12] LEGIFRANCE, Guide de legistique, 2012, available at http://legifrance.gouv.fr/Droit-francais/Guide-de-legistique/.

[LUT 16] LUTOFF C., CREUTIN J.-D., RUIN I. *et al.*, "Anticipating flash-floods: multi-scale aspects of the social response", *Journal of Hydrology,* vol. 541, pp. 626–635, 2016.

[MAR 97] MARTI R., LEPELLETIER T., "L'hydrologie de la crue de 1910 et autres grandes crues du bassin de la Seine", *La Houille Blanche*, vol. 8, pp. 33–39, 1997.

[MAR 10] MARCHI L., BORGA M., PRECISO E. *et al.*, "Characterisation of selected extreme flash floods in Europe and implications for flood risk management", *Journal of Hydrology*, vol. 394, nos 1–2, pp. 118–133, 2010.

[MAS 05] MASSOT M.-H., ORFEUIL J.-P., "La mobilité au quotidien, entre choix individuel et production sociale", *Cahiers internationaux de sociologie*, vol. 1, pp. 81–100, 2005.

[MON 16] MONDE (LE), "Inondations: la Seine-Maritime et l'Eure menacées, décrue amorcée à Paris", Le Monde.fr with AFP, 2016, available at http://www.lemonde.fr/planete/article/2016/06/04/intemperies-le-niveau-de-la-seine-s-est-stabilise-dans-la-nuit-de-samedi_4934610_3244.html.

[RUI 08a] RUIN I., CREUTIN J.-D., ANQUETIN S. *et al.*, "Human vulnerability to flash floods: addressing physical exposure and behavioural questions", *Flood Risk Management: Research and Practice Proceedings of the European Conference on Flood Risk Management Research into Practice (FLOODrisk 2008),* Oxford, UK, September 30–October 2 2008, pp. 1005–1012, 2008.

[RUI 08b] RUIN I., CREUTIN J.-D., ANQUETIN S. *et al.*, "Human exposure to flash floods – relation between flood parameters and human vulnerability during a storm of September 2002 in Southern France", *Journal of Hydrology*, vol. 361, nos 1–2, pp. 199–213, 2008.

[RUI 14] RUIN I., LUTOFF C., BOUDEVILLIAN B. *et al.*, "Social and hydrological responses to extreme precipitations: an interdisciplinary strategy for post-flood investigation", *Weather, Climate, and Society*, vol.6, no. 1, pp. 135–153, 2014.

[TAN 10] TANNIER C., VUIDEL G., FRANKHAUSER P. *et al.*, "Simulation fractale d'urbanisation. MUP-city, un modèle multi-échelle pour localiser de nouvelles implantations résidentielles", *Revue Internationale de Géomatique*, vol. 20, no. 3, pp. 303–329, 2010.

[THO 12] THOMAS M.-P., PATTARONI L., "Choix résidentiels et différenciation des modes de vie des familles de classes moyennes en Suisse", *Espaces et sociétés*, vol. 1, pp. 111–127, 2012.

[WAL 12] WALTERS B.B., "An event-based methodology for climate change and human–environment research. Geografisk Tidsskrift-Danish", *Journal of Geography*, vol. 112, no. 2, pp. 135–143, 2012.

[ZOC 10] ZOCCATELLI D., BORGA M., ZANON F. *et al.*, "Which rainfall spatial information for flash flood response modelling? A numerical investigation based on data from the Carpathian range, Romania", *Journal of Hydrology*, vol. 394, nos 1–2, pp. 148–161, 2010.

2

Social Effects of Floods on Mobilities: A Comparison Between Short-term and Long-term Perspectives

2.1. Introduction

Studying the social effects of flash floods is a difficult issue. The difficulty in analyzing these flash floods is that they are unique at every occurrence. They rarely happen in the same place, under the same circumstances and with the same intensity. Small drainage basins featuring this type of flood are extremely sensitive to the spatiotemporal distribution of rainfall[1]. While it is possible to count frequent occurrences of very intense rainfall on a departmental scale, it is more rare for this to take place several times on the same site (for instance, on the same rain gauge), and even less so on the group of rain gauges placed in the same watersheds, which triggers its violent reaction. Therefore, from one rainy episode to another, and even within the same episode, circumstances are highly changeable. This variability is reflected in the location of the affected watersheds, their reaction speed and even on the asynchronism between the flood's peaks [CRE 13]. While dangerous hydrometeorological coincidences are less frequent than intense rainfall, the probability for the same person to attend to these specific conjunctions twice in a lifetime is very small. In this way, every flash flood occurrence is highly specific and – over the average human life span – it is very rare to experience the same type of event twice. Just as an example, we could recall the episodes of 2010 and 2011 in the Draguignan area (Var, France), and those of 2009 and 2015, which took place in Mandelieu La Napoule (Alpes Maritimes, France). While in both cases the location of the rainfall and the affected drainage basins were quite similar, their dynamics

Chapter written by Séverine DURAND and Isabelle RUIN.

1 See Chapter 3.

and time frames were different, implying unprecedented experiences for the witnesses of these events, on the one hand; and incomparable consequences in terms of human impact, on the other hand.

Therefore, flash floods show an extreme degree of contingency. Now, the social sciences are feebly equipped for the study of contingency; so far, they have developed little work regarding the social effects of extreme events[2]. Only American[3] interactionist trends, or biographical approaches as well as the longitudinal dimension of social phenomena can account for these contingent aspects of the social world. This is the methodological orientation that we have chosen. Following in the steps of the theoretical work started by Grossetti [GRO 04] on the unpredictability of social situations, the aim here is to make sense of what emerges from a social point of view during the breakup situation that a flash flood implies. This question comprises not only the short-term nature of the event (understanding how the inhabitants react in the moment of the crisis), but also the long-term nature of a life trajectory (that is to say, particularly understanding how the issue of floods can influence residential choices). In order to observe the social dynamics generated by floods, we chose to work on two temporal scales: the short-term nature of the event and the long time scale of a human life. In this chapter, we intend to articulate the analyses of two dimensions: on the one hand, human decision making, based on the narratives of the actions undertaken during the flood (short-term perspective), and, on the other hand, the residential trajectories collected by the Life History Method [BER 97] (long-term perspective).

Regarding the short-term perspective, the intention was to retrace the causes that led individuals to become exposed to danger by examining the series of individual decisions and conditions that contributed to this situation. The main objective was to understand how floods interfere with the completion of a daily schedule and under what conditions people abandon their daily priorities so as to cope with hasty and risky changes in environmental circumstances. For this, we developed a method for conducting postflood investigations [RUI 14], in which the chronological guideline helped us to reconstruct individual stories by framing them against a temporality, a sociogeographical space and the network of interpersonal interactions imposed by the event. This method enabled us to explore the short lapses of time involved in the crisis (the temporal and spatial dimensions requiring measurements by the minute, per meter), while making it possible to identify or the larger scale factors that also influence decision-making processes. The first experiments of this method enabled

2 From a historical viewpoint, the social sciences were dedicated to the effects of structures, continuous processes and regularity [GRO 16], while they became increasingly detached from the hard sciences model [BES 10].

3 Namely through the notion of "career" [BEC 65] or the associated concept of "turning point" [HUG 96].

us to highlight the links between small and large scales, in particular regarding mobility choices. Moving in times of crisis was strongly conditioned by the daily needs associated with the implementation of a day-to-day schedule [RUI 07, RUI 14]. This schedule was made up of a series of subdaily tasks generally carried out in a small geographic area. It was the small-scale version of a life plan articulating family and professional projects forged over several decades and possibly involving residential mobilities to much larger areas[4]. This result later revealed the need to combine this short-term study with a longer term investigation.

Concering the long-term perspective, the same longitudinal approach of actions was retained, but applied to the extent of an entire human life, by concentrating on the reasons (or sequences of facts) for dwelling in a flood-prone area, staying there or leaving (especially after a flood). The life history method [BER 97] was refocused on the narratives of living practices from a chronological point of view (successive places of residence, motivations for change, etc.) and pragmatic practices (everyday practices deployed in the current neighborhood).

Both life project and daily schedule are limited by external events, and our analysis acknowledged this, whether limitations were linked to collective choices (land use, urban planning, the labor market, the real estate market, legislation, etc.) or to biophysical and technological constraints. This chapter aims to highlight the social convergences revealed by these two time scales. After giving the details about the methodology employed, we will present the results of our analysis of the social effects of floods on decisions and on individual actions in the short term, and then on residential trajectories (long term). Finally, we will rediscuss the forms of social logics convergences that emerge from this interscalar research, particularly *the force of everyday life*.

2.2. A common methodology deployed at two scales

Here, we offer detail on the methodologies we employed, respectively, so as to analyze the practices related to the event in the short term (section 2.2.1) as well as on the residential trajectories in the long term (section 2.2.2).

2.2.1. *Short-term perspective: analysis of postevent narratives*

The analysis of the actions carried out at the time of the crisis in response to the occurrence of a flash flood is based on semistructured post-event interviews[5]

4 These scale relationships will be further developed in Volume 2.

5 A type of interview conducted on the basis of a grid (a list of oriented questions), but where a large part is left for open discussion.

conducted with witnesses of the event. The technique is interesting in that it considers the real and specific situation as perceived by the people who experienced it as a point of departure [GRU 77]. By guiding the story chronologically, it is possible to finely reconstruct individual perceptions related to the evolution of environmental conditions, to social and conjunctural circumstances and to the specific emotions aroused throughout the event, as well as implemented protective measures [RUI 14]. Inspired by time geography[6], we have added the spatial dimension to the temporal one, in order to characterize the place of action, both in its spatial dimension (where the action takes place) as well as in its symbolic dimension (what this place represents for the person involved). Geolocation helps us to place the action in a hydrometeorological context observed by the sensor network (rain gauge, meteorological radar, liminimeter[7]). From the symbolic point of view, we endeavored to qualify the kind of place, whether it was fixed or mobile (house, work, vehicle, socialization arena, shelter, etc.), on the assumption that decisions made in the middle of an emergency situation are partly influenced by the constraints and representations associated with these places. For example, cars, as well as houses (in the sense of "home") generally provide a feeling of well-being and security that may sometimes be misleading in the advent of flash floods [DUB 06, RUI 07]. The interviews began with a question that aimed at setting in space and time the circumstances that led the witnesses to conclude that something abnormal, requiring their attention, was taking place; this helps to determine when the event actually started from their point of view. Starting from this moment, we retraced the sequence of events with them, making sure that for each action sequence (characterized by a unity of place and space), they would specify their perceptions concerning the external environment, their social interactions (face-to-face or distant interactions), their emotions and the measures implemented. Then, we represented these action narratives as individual event–time series, termed *actograms,* which helped us to perform a comparative analysis work (see an example in Appendix 1).

The analysis we offer here is based on a corpus of 63 interviews that document two different flash flood events: (1) 30 interviews conducted in 2003, 6 months after the dramatic floods of September 8 and 9, 2002, which devastated 297 of the 356 communes in the Gard department, resulting in 23 casualties and 1.2 billion euros of damage; (2) 33 interviews carried out in November 2010, 5 months after the floods

---

6 Time geography is a framework founded by the Swedish geographer Hägerstrand [HÄG 70] in the 1970s, which specifically considers the spatiotemporal embedding of geographic phenomena and processes. Time geography offers a conceptual framework for neatly describing and understanding the spatiotemporal arrangement and coordination of human behavior, as well as the dynamics of geographic spaces from the perspective of the individual/environment relationship [CHA 07].

7 Device for measuring the height of water in rivers.

of June 15, 2010, which ravaged 59 communes around Draguignan in the Var department, killing 26 people and producing damage equivalent to almost 1 billion euros [RUI 08, RUI 14]. In total, the sample was made up of 32 women, 28 men and two couples whose average age was 44, ranging from 15 to 86 years. The choice of communes for carrying out our research was made on the basis of geographical criteria in order to target three strongly damaged municipalities that were affected by different flood dynamics. Indeed, one of our hypotheses concerned the influence of the speed of the flood occurrence on the protective behavior of individuals. Thus, our research focused on areas prone to (1) a quick to very quick response of small catchments (smaller than 100 km^2) in the communes of Saint Hilaire D'Ozilhan and Remoulins on the Gard, Figanières and Draguignan on the Var; (2) a slower response of larger catchments (greater than 100 and smaller than 2000 km^2), comprising Draguignan and Trans-en-Provence on the Var, and Remoulins and Comps on the Gard. It is important to observe that Remoulins and Draguignan are set at the confluence of a small watershed (the Valliguière and the Riaille, respectively) and a large one (the Gardon and the Nartuby). The recruitment of the interviewees was done on the spot, starting from door to door in the flooded areas. Based on these first testimonials, we later interviewed as many people as possible who had been involved in the same flood events. Thanks to the comparison of testimonies, this "snowball" sampling technique enabled us to tap into the details of the decision-making process as well as the actions carried out during the crisis at a specific place. This technique also helped us to reduce the memory bias (inconsistency or inaccuracy), even if several studies showed that this was not so important in the case where the witness was personally involved, and the event was able to elicit emotional involvement [NOR 92, NEI 96, VAN 05].

2.2.2. *Long-term perspective: the analysis of life stories*

The study of residential trajectories through the stories of practices makes part of a broader socioethnographic study of the living in a flood-prone area[8]. The focus on residential trajectories proceeds by working on the factual elements that chronologically account for the residents' choices of places where to live. The research was carried out by comprehensive[9] interviews [KAU 96], conducted with inhabitants of three communes in the Gard department, the study area for the ANR Mobiclimex[10]

8 The methodology employed in this socioethnographic study is introduced in detail in Appendix 2.

9 See Appendix 2 for further detail on the methodology.

10 The Mobiclimex project was introduced in the Preface. The three communes mentioned are Collias, Sommières and Le Vigan. These communes were chosen because of the contrasts they show in terms of socioeconomic dynamics and of the local flood record. These study fields are described in Appendix 3.

project. While a socioethnographic study makes it possible to highlight the specificities of each life space taken into consideration, refocusing the analysis on residential trajectories deals with the individuals on the same level: that of the inhabitant of Gard, a department strongly exposed to flash floods. The analysis presented below is based on 27 interviews conducted with inhabitants (11 men and 10 women) or couples of inhabitants (six interviews). The sampling of the interviews, bearing in mind the balance between age criteria, location of the house in a flood-prone area or not, prior flood experience as well as seniority in the territory. In the life stories interviews, the resident was invited to consider his or her experience through a filter: that of the residential trajectory. The initial common framing favored the comparability of different stories. It implied inviting the respondent to retell his personal history in relation to his habitat, within a longitudinal perspective. The respondent was invited to enumerate his successive places of residence and to detail their specificities (location, type, advantages, disadvantages, etc.). Every change in the place of residence was questioned so as to value the events that led the interviewee to move. The interviews were recorded and the narrative material was fully transcribed for analysis. In this project, the focus on residential trajectories made it possible to build *individual chronological frescoes* that favored a representation of the successive places of residence in the light of potentially influential events (see an example in Appendix 4). Thus, the trajectory of each inhabitant is represented chronologically on this fresco – which we call *inhabitant chronologies* – highlighting the changes in place of residence in relation to personal events (marriage, divorce, birth of children, etc.), professional events (training, hiring, job loss, retirement), or external events, particularly floods. A comparative work on the resident chronologies constructed in this way finally aimed at identifying the common elements, the regularities or the underlying logic that could be identified in the face of the flood considered in view of the whole of the life's events of the 33 total interviewed inhabitants (17 men, 16 women).

2.3. Social dynamics in the face of extreme events

In this section, we will introduce the results of our analysis of the social effects of floods on individual decisions and actions in the short term, and then on residential trajectories (long term).

2.3.1. *Setting the event in time: when the flood imposes itself on the inhabitants*

Research describing and explaining decision making and behavior in response to environmental hazards and natural disasters has been prolific since the 1970s. It arises from different disciplinary and interdisciplinary streams including sociology, psychology and geography, depending on whether we are interested in collective responses, cognitive processes at stake at the individual level, or the interactions between physical factors and human factors. Our study heavily relies on the protective action decision model (PADM) developed by Lindell and Perry [LIN 03, LIN 12] to study the context that influences the adoption of protective actions in the middle of an emergency situation. This model brings forward several factors that trigger decision-making processes and protective measures. Here, we will refer to these factors as "precursors"[11], in that they are related to the different contexts of the action situation. Among these, the environmental context is related to the physical processes that determine the event's speed, its magnitude, the extent and duration of the disruption, as well as the presence (or absence) of signals from which the observer may detect the emergence of a threat. Then, the social context is based on the observation and interpretation of the behavior of others. The context of the alert, whether an official message or not, provides details about the characteristics of the content and the transmission of information prior to the event. Finally, the personal characteristics of the alerted individual should also to be taken into consideration: the person's physical, psychomotor and cognitive capabilities, their social capital condition to access the alert, as well as the ability to react to the imminence of the threat. According to Lindell and Perry, these factors feed cognitive processes in three stages. (1) Predominantly automatic predecisive (unconscious) processes–determined by preconceptions, competing demands for attention and the intrusive character of information–make it possible (or hinder) awareness about a possible threat. (2) The perception of the threat, its potential impact and the ability to become protected against it are anchored in long-term mental patterns or representations via direct personal experience, learning, peer exchanges or passive exposure to media information. (3) The decision to seek safety arises from the identification of the threat, the personalization of the risk and its consequences, and from the evaluation of alternative courses of action, mainly by weighing up either implementing protection activities or carrying on with "normal" everyday activities. In this process, it is necessary to establish steps for research or for confirming information from a variety of sources, in order to better understand the pros and cons of the situation, and to optimize the response. This process may take time and diminish the relevance of a solution that could have been appropriate had it been implemented without delay. In the same way as seeking information, seeking to protect property may inadvertently delay the protection of individuals. Finally, the actual implementation of a protective behavior is often subject to the contingencies of the situation, to potential physical, social or material obstacles or, on the contrary, to opportunities arising in the heat of the moment.

11 This notion of precursors is developed and discussed in Chapter 8.

On the basis of this explanatory model, our interviews shed light on many of the factors and processes described by Lindell and Perry. First, we considered the way in which the specific characteristics of flash floods (sudden triggering of the event, difficulty in predicting the location and time of the impact, asynchronous flood peaks[12], etc.), as well as a variety of associated perceptual experiences may have influenced how individuals interpret the situation. Then, we focused on the practices implemented in responding to the urgency of these situations, and the motives underlying these behavioral choices.

2.3.1.1. *Representations questioned by the event*

The testimonies of different flash flood episodes make it possible to partly retrace the representations that individuals make of these phenomena and the dangers associated with them. The speed of the rise in water, the heights attained as well as the location of the flows and the mechanisms involved, all work as surprise factors for questioning the event. Let us explore these aspects in detail.

Inconceivable events

In communes located in very small drainage basins, as is the case in Figanières, Saint Hilaire d'Ozilhan and Remoulins (Valliguière catchment), the word that appears most often in testimonies is *wave*, so as to describe the sudden influx of water that people face. The word in itself summarizes the speed and the water height that imprints the mind.

> "In 10 minutes, a quarter of an hour we had 1.60 meters of water. [...] rain did not stop; it was uninterrupted since the previous night, but really heavy, heavy, heavy, non stop. We had never seen that. [...] We reckoned we should not go out, otherwise, we would be soaked, but we never thought things would happen like that. [...] Truth be told, we know the place, we know where, when there are weeks of rain, when we will have our feet in the water. But we never imagined that a wave like this could happen. [...] But well, all in all, we never think that, we think about the floods in the plains, but here we never think a flood will cross a river and reach us" [Trader and inhabitant of Figanières, 51, native of the Mediterranean region].

As several witnesses pointed out, this type of phenomenon goes beyond imagination and understanding, far from the classic images of floods in the plains that spread slowly until it submerges a large area.

12 This particular feature of flash floods is described in Chapter 8.

> "It was my first flood experience. It was traumatic; water rose and never stopped. I am from the Somme, where water is stagnant, this was extremely brutal. In 2 or 3 hours, while we were here, there was 2.50 meters of water" [Inhabitant of Remoulins who had been living there for 7 months at the time of the flood, 34 years old, teacher].

Local flood scenes are all the more incomprehensible since in general, there is no real visible river nearby, but rather what is locally known as "valleys", most frequently dry, small draining surfaces that is hard to imagine that can collect so much water.

> "I said to myself: the small valleys are probably going to overflow, embankments and roadsides will collapse with the amount of water, but nothing else. I said to myself: we are going to be on call, because sometimes when the soil is very, very wet, there are pines that come off. [...] Here, we have two Roman bridges which were flooded. The Roman bridges hadn't been affected so far. Yes, in one of the valleys. We have a Roman bridge that is 2000 years old which has disappeared" [Inhabitant and locally elected representative of Figanières, 63 years old, retired].

Between the flooding of the valleys and the superficial drainage favored by impervious surfaces and the roads, water sometimes takes unexpected paths, which increases the level of surprise even more, and does not make the understanding of the situation any easier.

> "But, honestly, the flood that came along, it was illogical. All of a sudden, a calculation was made. But it was impossible for 1.80 meters of water to be there. I was looking out the door. I saw it was not flooded. That is to say, it was not yet the complete catastrophe. I said to myself: this is not possible, the water has risen not only there, but it is everywhere. Nothing made sense, it was completely illogical" [Artisan in Figanières, 40 years old].

In the end, it would be rare for flash flood witnesses like those who lived in Figanières, Saint Hilaire d'Ozilhan and Remoulins, to be not surprised by the brutality of the events and the violence of their impact. Even if the extraordinary character of the rain is often mentioned, and sometimes confirms the announcement of orange or red meteorological alerts, this type of rain is rarely interpreted as a signal of an imminent and dangerous flood.

A great difficulty in making sense of the situation and in adapting one's behavior accordingly

Numerous testimonies reveal that alerts (orange vigilance) and warnings (red vigilance) were announced by Météo France before the beginning of the flood. Nevertheless, in none of the testimonies were alerts actually taken into account for beginning with the protection, either due to a lack of confidence in the forecasts, or because people considered that they did not apply in their case. It is also true that the circumstances leading to a flood can vary greatly. In communes drained by a small watershed, an active rain–storm system located on the catchment suffices to cause a rapid rise in water levels. In the case of a large basin, as in the case of Comps, heavy, widespread and stationary rain is required for the entire basin (> 1000 km^2) to react to the watercourse. In the latter case, there is a time lag between rain and floods: without a sight of what is happening upstream, it is difficult to predict what will come. The testimony of a locally elected Comps representative illustrates this problem.

> "The red alerts of Météo-France, they worried me and they did not. I was worried about work, in the worst of cases, if there was a big storm, I would not go to work, but concerning watercourses, we know that there is a flood warning chain. The weather forecast announces impacts and we all agree, we know that there will be water, but water can fall locally and not up here, so the flood is not linked to meteorology. It is a local phenomenon, unfortunately it was all over the watershed, it was unparalleled. A storm does not necessarily mean a flood" [Comps locally elected representative and resident for 25 years, 52 years old].

Meteorological warnings are aimed at the spatial scale of the department, bringing attention to not only the occurrence of stormy rains but potentially more or less localized floods. Consequently, they may seem inadequate for a large part of the population of the department faced with phenomena that most often occur locally. Thus, the spatial and temporal uncertainty of the forecast works as a source of doubt for the inhabitants and introduces difficulty in personalizing risk.

> "Météo France said that we were on a red alert on Sunday night, but well, considering that they are often mistaken, we did not trust Météo France too much. There were alert maps. Were we to see them again, now we would be a little bit more shocked. It is true that a red alert is sometimes a bit exaggerated, it happened more than once that they were wrong, and as they are more frequently wrong than right. We do not pay so much attention to them any longer, but well, apparently they were not wrong on that one. We did not make the mental

> connection between Météo France and the floods" [Native and inhabitant of Comps, 32 years old, housewife].

Even when the level of vigilance is red (the maximum level), locally contradictory information such as the closure of schools but not of the local day care center can increase confusion. Let us take a look at the following testimony:

> "Météo France, frankly, they did not foresee it anyway. At some point, there was an alert, orange then red, but not for us, it was not for Remoulins. I remember hearing on the radio that high schools would be closed but in my working environment, no one called to say close the day care" [Inhabitant of Remoulins, 46 years old, employed].

In communes such as Remoulins, located at the confluence of a small and a large drainage basin, it is necessary to cope with various flood dynamics and multiple flood peaks. This was the case in September 2002: the Valliguière basin quickly reacted to the rain, and it was followed several hours later by the flood of the Gardon, even if the rain had stopped: an unexpected situation even for the managers of the crisis.

> "When I went to look for my wife on foot because it was impossible to drive over, it was 3:30 pm; the firemen told us not to worry since rain would stop at 4 pm and be over. While it is true that the rain stopped, it was from that point that water started rising" [Inhabitant of Remoulins, had been living in the commune for 10 years at the time of the flood, 34 years old, employed].

As the last testimony shows, the main difficulty lies in picturing a general idea of the situation so as to consider what can happen due not only to local circumstances, but also to the more distant context. Perception and the anticipation of environmental circumstances are necessary for envisioning safeguard behavior. While this may not always be evident even when visual cues are available, it is even less obvious at a distance, especially when considering a driving route.

> "In order to be aware of what is happening, it is imperative to have an overview of an area or a department, everyone could acknowledge the measure of what was happening at home, but did not have the big picture. So, we were not able to measure the extent of the problem, we are always short-sighted. [...] I would not still change my driving routes, because if we take a look at Bagnols, the Cèze overflowed, who could have believed the Cèze was going to surpass the bridge that goes from Bagnols to Avignon, this was absolutely unpredictable. Harmless little streams became monsters, so this is something

> unimaginable before we see it, so saying I will not take this way, instead of that one… Anything can happen anywhere" [Inhabitant of Saint Hilaire d'Ozilhan, retired, 60 years old, former inhabitant of Nîmes during the 1988 flood].

From one flood to another: the ambiguous role of flood memory transmission

In the social science literature on risk, the experience of similar past events and the proximity to the source of risk (in this case, the river) are often considered as factors that increase perception and risk personalization [WEI 89, LIN 12]. Some witnesses of the floods of 2002 in the Gard were able to compare them with the floods of 1958, which they had also experienced. Despite several more recent flooding events in Remoulins, the one of 1958 was the only one to reach inhabited areas before the flood of 2002. The commune of Comps, on the other hand, had never experienced flooding since the construction of a protective dike in 1928 until the episode of 2002. These circumstances may explain why several inhabitants of Remoulins and Comps underestimated the flood of 2002, thinking that it could not possibly exceed their previous markers (floods of 1958 or 1928).

> "Ah, in 1958, I was still living with my parents, we saw it... on the first floor, there was only one room which was flooded, but this time it went up in the other rooms too, one more meter at least. [...] the flood had arrived quickly in 1958 too, but this time, we never thought it would arrive here" [Inhabitant of Remoulins, 60 years old, retired].

Experience combined with an in-depth knowledge of the long history of flooding in the area may act as a better ally for understanding the processes and providing a better representation of the associated perils.

> "When we left abruptly from here, there was water on the road. We could not leave, we went across the fields when we saw that the flood could overflow the dike. [...] We left because I was aware that it was dangerous, from my experience and what I studied about ancient history, destroyed churches, the strength of water. What you have to imagine is that a cubic meter of water is equivalent to a ton, when it moves at 5,000 meters^3 of water per second, imagine how strong it becomes; 5,000 tons per second, nothing can resist that" [Inhabitant and native of Comps (where his family had settled for generations), 82 years old, retired farmer, commune historian].

In light of these varied testimonies, it is clear that to the local villagers, flash floods appear as improbable, unexpected and unimaginable events. Not knowing what to expect, or on the contrary, expectations based on representations of a slow-spreading classical flood, makes it difficult to make sense of local situations whose dynamics are very different and do not correspond to the unique signal given by the rain–flood alert from the authorities. In these circumstances, it is not surprising that responses to alerts may take time. In section 2.3.1.2, we will try to understand how this time span is employed and how it contributes to the likelihood of anticipating protection.

2.3.1.2. *From alert to protection*

While the extracts from previous testimonies enlightened us on the preconceptions associated with the phenomenon of flash floods, we now discuss the cognitive processes conducive to a state of alert and the action of implementing protective measures in the face of rising water levels.

The necessary accumulation of evidence for detecting the abnormality of a situation

As summarized by Mileti [MIL 95], many studies have shown that a response to an alert was the result of a social and cognitive process involving the following stages: listening, comprehension, belief, personalization and information confirmation. The following testimony is a good example of this sequence:

> "I was here, so I ran the shop all day, so I was working. We will say that around 5pm, a customer warned me against the rain, knowing that it was raining a lot, but we did not worry a great deal about that [...]: "We cannot go in, the way is blocked, it is flooded". I looked at him and said, "Are you sure? Maybe there are men at work?" He replied, "No, it is blocked, it is completely flooded, I have already struggled to get back here." Then, I started worrying and called my husband. I said, "Look, it's weird." to which he replied "He must have been wrong", because it is true that we did not believe too much and let's say, maybe two minutes later, another customer came in and said: "The water is at the level of the bridge, that is to say opposite to the town hall" [...] "The water is up, there is still almost a meter left, before the water washes all over the steps". I looked at her and said "It is not possible", and I thought she must have drunk some alcohol. So, again I called my husband and I told him: "Listen, go and see". So he went to see and said, "Yes, it's true. Listen, shut the shop down for two minutes, and go see because it's impressive." He said to me, "I think you will not believe it." So, I closed the shop and went to see and then that is when I saw the water. What was left? There was barely a meter left before the water overflowed and got to the road [...]

> the bridge was shaking uncontrollably, so as we went up on the bridge, we stepped back, because it would surely be unable to hold, so fiercely it moved [Merchant and inhabitant of Trans-en-Provence, 27 years old].

This testimony shows to what extent the process can be iterative and slow (particularly when the response time of the basin allows it), requiring interactions not only between several people (including confirmation with a person of trust, such as the husband in this case), but also visual confirmation by going in person to check the state of the river and feel the "shaking of the bridge".

At a cross-roads: the decision on whether to stick to daily routine or to deal with the emergency

Another reason explaining late behavioral responses is related to the desire to continue with everyday activities. As we will clearly see in the following testimony, our daily activities may trick our attention, preventing us from seriously considering any piece of information that emerges from the environment.

> "In fact, my wife was at work and I was at work too, so we didn't notice. We could see that it was raining, and we said to ourselves: We will carry on with our daily routine" [Inhabitant of Draguignan, territorial agent in Figanière, 28 years old].

As we will see in the following two testimonies, going to work appears as a strong source of motivation for ignoring the circumstances.

> "At around 7 pm, it started raining heavily, when I left, my neighbors were already making holes in their walls. As for me, my place was not flooded yet, since water was flowing in front of us, so I went to work. I should not have gone however, but well, I never thought... The roads were already flooded but the road to Castillon, there is a place where the road is flooded only a little. When I went that way, there was water already, a little less than ten centimeters, I said "it's okay, it's a storm, it will calm down" [Inhabitant of Saint Hilaire d'Ozilhan for 12 years at the time of the event, 43 years old, employee].

> "On Monday morning, I went to work normally, I took the same route as usual and at around 12 o'clock, when I came back home, there was water in the garden at the back, below. The garden leads to the Vigière, it is about six meters above the Vigière. There, I started to worry. I warned the day care and told them it was necessary to close because the water was rising and it was disturbing. The director did

> not want to close. At 1 pm I went back there by car, we could still move around, when I arrived, they were closing the day care; the parents had come to pick up their children. I immediately went back home, the road was still not flooded, but it was completely blocked at 3 pm [Inhabitant of Remoulins, 46 years old, day care employee].

In these two cases, it seems that the problem is not the lack of attention, because the two narratives show a state of alert related to unusual and disturbing circumstances caused by the floods. Here, the professional activity appears rather as a (moral or economic) constraint, difficult to dismiss. In the last testimony, the employer's intervention reinforces the sense of constraint, which ends up taking over awareness.

Protective measures: between a thoughtful act and a reflex movement

Depending on the flood dynamics related to the size of the catchments, the situations people face differ greatly. In the case of very small basin, the reaction must be fast, because the response time to rain peaks barely leaves about 10 min for protecting oneself. The following testimony underlines the fact that protection is more a reflex movement than a well-considered act.

> "I was drying a little water that had come in. It was barely one centimeter high. It was serious because the cars in front were covered up until the level of the license plates. And then, little pieces of wood started coming in. More or less, I was coping with the situation, when suddenly I heard some crackling [...] I heard "crack", it was the door, starting to give way. I was looking in that direction when I saw 1.60 meters of water, which was behind. There, I lost control. It was no longer manageable. I cut the electricity. I had about twenty seconds to stupidly try to put things in a safe place, and then the door exploded. All of a sudden, I had 1.80 meters of water flowing inside. [...] I had been managing the situation that came from one direction and all of a sudden, behind me, a new situation which was no longer manageable. Then I tried to think, but there was nothing to ponder. In any case, the door had exploded. I was right there. I jumped outside. But honestly, the impression I had was that I was in control of the situation and then, at some point... well, I had been able to control my little flood, but then I could control nothing at all. [...] We no longer knew where to be safe anymore. We did not know where to go to protect ourselves [Artisan in Figanières, 40 years old, inhabitant of Calas].

On the contrary, when the size of the basin leaves several hours for reaction, the inhabitants have enough time to implement an analytical strategy of the situation, therefore more progressive and thoughtful protection.

> "The first thing we did was to put our cars in the shelter, without really thinking that the water would come in our direction. Then we saw the water arrive via the road. We had put pebble stones on the entrance. We saw that they were hit by the water, it was around 7 pm. We saw the water arrive in the garden, so we started going back home. We put protective earth sacks in front of our house, in front of the door, in the toilets and when the water started to arrive in the house, we took buckets and we were pretty optimistic about draining the water, when we realized that we were submerged, so we went up the stairs. I don't know why, we had decided that the water would stop at one meter, so we all climbed above one meter, but in fact, water climbed 1.50 meters. We said, we can do nothing else, we have to wait for this to pass, so we went to occupy the first floor with our dog, and we waited all night long" [Inhabitant of Comps, 62 years old, retired, native of the commune].

In both cases, the first reflex is to offer resistance and to control the situation by blocking or channeling the passage of water[13] in order to protect its normal functioning. But in the end, acceptance becomes the only possible solution.

In this section, we have highlighted the importance of the spatiotemporal context for decision making. The temporal pressure and the stress imposed by the response speed of the basins, or the context associated with the time of day or the types of places in which the action develops, play a key role in decision making choices and constraints in terms of protective measures. Nevertheless, as we will see in section 2.3.1.3, these are not the only contexts.

2.3.1.3. *The importance of context in decision making*

As a complement to spatiotemporal factors, factors related to the social circumstances of decision making should also be taken into consideration. That which repeatedly appears as a factor of caution (both for men and women) is the fact of being in the presence of their children and being concerned for their safety.

> "In danger? I was scared for my daughter. I left with her in the car because if I had been alone, I would have tried to return to

13 This resistance reflex is discussed in Chapter 8.

> Draguignan. But with my girl in the car, I was afraid of getting stuck, of not being able to feed her, to give her something to drink. It was cold, wet, all that. All alone, yes. If I had been alone, I would have gone to Draguignan. I would have taken another route and I would have found out after that we could more or less circumvent it. I could have tried. [...] but I think that fortunately, there was my daughter, because I might have done something wrong at that time" [Inhabitant of Draguignan, territorial agent in Figanière, 28 years old].

Responsibility toward children seems to guide the most cautious behavior. Previous flooding experience may also favor a context of prudence, as the following excerpt concerning the flood of Nîmes in 1988, reveals:

> "I saw the water coming up to our house, and there were mattresses floating, there was a bit of a current. I said to myself, I don't care about anything, I will take my children and leave, we can buy everything again, but if your children are gone, you will be dead. I was not afraid for the house, I didn't care, I took the papers. The water rose from 3 pm to 6 pm [...]. I went to the neighbor's house with my children, near the car park above Champion, everyone was there because it's a high place" [Inhabitant of Remoulins for seven years at the time of the event of 2002, 39 years old, housewife, and victim of the Nîmes flood in 1988].

This decision-making context (past experience combined with the presence of people in charge) exceeds the short term of the flood; it should rather be understood in the long term of a life trajectory. The interdependence of short-term and long-term actions motivated the interscale perspective of the present analysis. Section 2.3.2 will be devoted to the long-term perspective of residential trajectories.

2.3.2. *The events in perspective: when life choices are more determining factors than the flood itself*

In common thought, the choice to live in an area exposed to natural risks sounds counterintuitive or even illogical. Nevertheless, several case studies showed up to what extent the question of risk does not constitute a sufficient reason for moving home or for refusing to settle [DE 03, FAB 03]. Illogicality as an *a priori* for settling down or for deciding to remain in a risky zone falls within the supremacy of rational logics that, after investigation, is not consistent for explaining the residential choices of individuals. In fact, the postulate of the hedonic[14] price method has been widely questioned in terms of residential choice [GUE 06]. The study of residential chronologies showed that for the 27 cases under study, in the great majority of cases – except for some cases that we will analyze later – resident trajectories were not strongly influenced by the question of flooding. As we mentioned earlier, in the Gard region, the choice of a place to live results from complex compromises that involve all the dimensions of existence [AUT 10]. Compromises do not simply refer to the application of a utility calculation under constraint [GRA 10]. Understanding residential choices does not only involve paying attention to people's resources and constraints, but also taking into account the contextual elements that model their decisions, the motives they expose as well as the interactions with others, and their personal and family history. It is a question of "studying situational choices" [GRA 10]. Residential behavior depends on socially embedded values, which are closely linked to the environment in which the person grew up [BON 99]; exactly the material we are going to study in depth. Accounting for the significance that inhabitants attribute to living in a flash flood-prone area, we will discuss the reasons given for their choice of place of residence (1), the explanations they provide concerning the fact of staying there despite the risk (2) and finally, the reasons exposed by those who decided to leave (3).

2.3.2.1. *The motivations for living there*

The analysis of life stories in terms of habitat makes it possible to highlight three main criteria for explaining the reasons that motivate the choice of location. In ascending order of relative importance we can mention: love for the environment, lifestyle choice and attachment to the place and to those who live there[15].

Love for the setting and the characteristics of that special environment

> "We love our region, we love our environment".
> [Native inhabitant, around 30 years old, business person, Le Vigan]

The natural history of Gard features an attractive landscape, combining the plain, the mountain and the gorges, which makes it ideal for outdoor activities (hiking, mountain biking, canoeing). This natural heritage is combined with a remarkable architectural heritage[16], which has turned the department into a popular tourist

14 A method according to which individuals might be rationally motivated in their choices and might arbitrate between the objective characteristics of accommodation and the specificities of the environment.

15 The importance of transmission of family real estate property – which we shall develop in another chapter (see Volume 2) – appears as the main influencing factor on these criteria.

16 Gallo-Roman remains are numerous and many exceptional books have made them famous (among which we can mention the widely known *Pont du Gard*). A great number of

attraction[17]. The living environment is commonly known for being particularly pleasant. For a long time, it was set aside in the study of residential trajectories, but nowadays the quality of the environment is recognized as an important factor [GUE 06]. Several studies have shown that residential choices are increasingly sensitive to the surrounding landscape [DOB 95, COL 98, MAR 99]. This importance was evident in the testimonies we sampled among the Gard inhabitants we interviewed. The love for the living environment as well as its green nature was not only highlighted by the natives from Gard, but also by those coming from other regions, whose to live in the Gard region was largely influenced by its natural environment.

> "You go down on the edge of the Gardon when there is no one there, and one evening during the week, when the temperature is 25 degrees in September, you see two beavers crossing the river [...] you have an eagle that flies above them, and ... very silent, it's really nice... you're there thinking: wow!". [Inhabitant of Collias, about 40 years old, native from another region, moved to the commune five years ago.]

The narratives about the living environment, most often conducted in a positive light, are associated with those related to a lifestyle that meets the expectations of the inhabitants.

Choosing a lifestyle

Research on residential trajectories has captured the importance of considering the variables related to lifestyle that explain the choices made by residents (this implies going beyond traditional sociodemographic variables) [BEL 68, BRU 94, PIN 02]. The stories collected here focus on the importance of this lifestyle choice considered pleasant in everyday life. The appreciation of this lifestyle is mainly associated with the concept of accessibility, linked to everyday practicality.

> "The quality of life is wonderful here [...] you want to go to Montpellier [...] 25 minutes and you are there. You have Nîmes next door, the Cévennes an hour and a half from here to visit the mushrooms, all that, I love. The sea is near and we can go fishing, for

---

communes in Gard have been classified as "the most beautiful communes in France" and the Roman heritage of the *Préfecture du département (Nîmes)* was classified as World UNESCO Heritage in 2016.

17 The Gard Bridge itself draws more than 1 million visitors per year. The annual return associated with the Tourism industry in the department is calculated at around 970 million euros (Source: common research conducted between Gard's agency for development and tourist reservations (*Agence de Développement et de Réservation Touristiques*, ADRT) and Gard's *Chambre de commerce et d'industrie,* CCI), 2015.

> a quiet swim in the ponds, great stuff. It is a beautiful landscape. In fact, you have everything, the fishmonger, the butcher, there are quality traders [...] there is a greengrocer there is... there is everything" [Inhabitant of Sommières, around 45 years old, arrived from another French region five years ago].

The narratives of Gard's inhabitants confirmed that residential arbitration is closely linked to the desire of being able to endorse a satisfying lifestyle, from a functional point of view [THO 09]. The choice to live here is justified by organizational issues, closely linked to family and friendly networks of solidarity – particularly for parents. The presence of a local solidarity network for day-to-day family organization (for example for picking up children from school) may be mainly linked to the value of local sociability. Indeed, this element was emphasized by the inhabitants; local sociability that can be described as what we may call a rural community [HAN 83].

> "Here there is still a small group... [...] that's what we appreciate, that's why we stay here, besides [...]. We know that there is someone nearby we can rely on" [Native from Collias, around 60 years old].

Attributing value to sociability is associated with a feeling of belonging to a local community, which often arises from personal and family history. Previous studies have demonstrated the influence of reference places (linked to parents, grandparents) and founding spaces (past residences) on residential trajectories [BON 99]. This influence is verified in the sample we interviewed. Such an influence is related to the love for the environment, but even more to these attachment criterion.

Attachment to the place and to its people

The question of attachment to the living space is what our interviewees emphasized the most. In fact, in the majority of cases, this issue appeared from the very beginning of the interview. Most of the times, attachment appeared in two intertwined forms: an attachment to the place and an attachment to the loved ones who shared the same living space. It seems that this question is embedded within the question about the origins.

> "Here, we are well, we are from Collias. I gave the choice to my children, I told them: either it's Nîmes, or it's Collias, they chose Collias". [Inhabitant of Collias, about 40 years old, returned to her home community five years ago.]

The question about the origin of the inhabitants (placing oneself in relation to the historical location of the family) was mentioned both by native inhabitants as well as those more recently settled. Regardless of categories, this shared reference reveals the relevance of the question in the Gard milieu. Not only do natives pride themselves on the fact of coming from this region, but also the more recently settled inhabitants claim their roots in the area.

> "I'm from here. My grandfather was from Collias, so you see. [...] All in all, I came back home. [Inhabitant of Collias, around 50 years old, who had returned to her commune ten years ago.]

> "I feel as if I had been born here in Sommières, [...]. Yes, I am attached to this place". [Inhabitant of Sommières, around 45 years old, who moved here from another commune in the department more than ten years ago.]

Even clearer than in Chardonnel *et al.* [CHA 15], it was illustrated in the neighboring department of Ardèche (sharing a similar landscape and socio-economic characteristics), where the reasons for choosing a place of residence are explained by the Gard inhabitants primarily based on a feeling of attachment to places and to others which is featured in the autobiographical narrative as well as in the fashioning of people's identities.

In section 2.3.2.2, we will try to understand the ways in which a flood may (or may not) interfere with the perception of the qualities recognized in the living space.

2.3.2.2. *The reasons for staying despite the floods and their risk of occurrence*

> "We do not even discuss moving away... Then, there are those who are used to the place [...] and who take preventive measures, after which they arrange their houses differently" [Expert resident (insurance consultant, in charge of files concerning flooded families), Remoulins].

The counterweight of the other reasons for staying there

For the majority of the inhabitants we met, whether their dwelling was set in a flood-prone area or not, the fact of being exposed to floods did not appear in the narratives of the inhabitants as a sufficient criterion to justify leaving the territory. Some residents in homes safe from floods even said that they would have liked to be closer to the river, despite the risk factor.

> "I have a lot at the edge of the Gardon, which is I cannot construct on and yet, if I could build, I would build, honestly". [Inhabitant of

Collias, about 40 years old, had come back to the commune five years previously.]

While some of them were able to choose to leave after a flood (as we will see in a case in the following section), this question does not predominate in the narratives of the interviewees (neither directly nor reported). For the majority of them, this was not a sufficient reason for leaving the place.

Interviewer: Have you ever wondered about that?

"No. [...] Had I been a bit more cautious, perhaps, I would never have lived there. But I always liked it [...] We are still a little away from... the city. It is calm here. The houses are nice. No. I never wanted to leave". [Native from Le Vigan, had come back to the commune 30 years ago, evacuated in 2014.]

When asked about the reasons that led them to this position, the inhabitants mentioned the appreciation criteria described in the previous section as counterbalancing their choice. When questions were pushed beyond these criteria, several other positive factors were still mentioned, depending on the individual. But above all, the importance of risk was downplayed.

A degree of severity perceived as rare

In order to make the case for this relativization, primarily it was the rarity of occurrences that was highlighted, such is their exceptional character, compared to other assessment criteria. While from an external point of view the regularity of the floods in the region may appear ongoing, for those who live in the territory, major floods remain minimal, and even exceptional. Those who perceive the regularity of the floods are interested in the whole region, which is surely regularly affected. Being interested in the inhabitants means taking a much finer scale into account, that of the community, even of the district or the abode. Given the type of hydrometeorological phenomenon we are referring to, events are rare on this scale[18]. The probability of events affecting the same house several times in a row is low, despite the fact that the occurrence of this type of phenomenon is widespread on the scale of Gard as a whole.

The inhabitant whose testimony we will now quote is native to the region, and her childhood memories were marked by the flooding of her grandmother's house by the river. However, flooding was not a criterion taken into account in her choice of housing. As from the first year when she settled, her garden was flooded (and

18 The question of relative scales is analyzed in Chapter 1.

again shortly before our interview). The witness argued that this had not motivated her to leave her house, where she feels utterly comfortable.

> "I never wondered about that. I asked no one. I did not go to the town hall to see the map of this, or that, after all [...] it was back in... in 1982 that I arrived. Well, in 1983, there was a flood, too. Finally, the water [...] rose high, very close to the sliding doors. I had not been back home between noon and two, I had not been able to go back because my engine had flooded in the morning. [...] I mean... we had been there for a year. But we never said: hey, let's leave, because it gets flooded here. [...] We had been there for years and it had not been flooded. It can take around four years for this place to get flooded... well, there's the Coularou [small stream] that rose a little bit, but... but not... not dangerously. [...] And we never said, "We... we'll leave".
>
> **Interviewer: Isn't that enough reason?**
>
> "No. I mean..., no. I never wanted to leave. No. And especially considering that water does not rise every year. It can take three to four years for us to have floods here. No, I'm not saying that if there had been a flood like we had this year [Coularou flood, September 2014, Le Vigan], every year... maybe I would have wanted to leave" [Native from Le Vigan, moved back to the commune more than 30 years earlier, evacuated in 2014].

A large majority of the inhabitants we met underlined the exceptional nature of major floods. This exceptional nature of floods – in a region that is nonetheless highly exposed – was raised almost unanimously by the inhabitants. The fact that flash floods are frequent at departmental scale, but that two similar occurrences to take place at the same point in the territory are rare, explains this representation of risk. Discourses alternate in a rather ambivalent way between the confirmation of the events' recurrence – leading to a reflection on the irreversible nature of a flood – and the observations about the "exceptional" character of the last local event. This ambivalence sometimes appears within the same argument.

> "No, it was rather an exceptional occurrence [in reference to the last flood in 2014, during the autumn preceding the interview]. That is to say... Every year, when we reach this period, September–October even until mid-December, we are in a period considered relatively risky" [Inhabitant of Remoulins, had arrived in the commune from another region ten years ago].

Thus, a majority of the inhabitants downplayed the flood risk concerning them directly. As it has already been shown in other areas [DUR 14], the fact that concern[19] with the flood mainly focuses on the criterion of flooding the interior of a house contributes to explain risk relativization. Cases where the water completely flooded (or could have flooded) a home (or even totally destroyed it in some cases) are much rarer than the cases where the damage is moderate.

Staying at all costs: the attachment to the place and to its people

Among the six inhabitants we interviewed who had been faced with a flood in their current home, three of them defended their desire to remain at this location despite all odds. Two people (a couple) ended up giving up for economic reasons (the last case mentioned above) and another one decided to stay for the moment, but admitted feeling anxious every autumn and dreamt of leaving some day. Among the three inhabitants who decided to stay (or would have liked to), one of them, whose professional establishment is exposed, enforced important measures that now allow him to "manage floods".

> "I have my own criteria. I have my rating, I know that when there are 3.23 meters in the Vidourle, there will be water in my restaurant [...] When I know that there will be a "Cévenol-type" flash flood, I remove everything before hand, I close earlier. [...] [And after the flood recedes] Everything is prepared [...] I open all the doors and I start to scrape everything so that the silt does not settle on the floor. Because that's it, the worst is the silt. [...] A jet of water, everything is already set, clean, water everywhere, I put the fans at top speed at a cold temperature, then at a hot temperature, otherwise, it creates moisture".
>
> **You are very well organized!**
>
> "Oh yes, it is a great deal... full of fans, also the plugs, which have to dry thoroughly inside. The next day, we reconnect everything and it all starts anew. [...] We can manage where it will have an impact, and all, roughly, by following the event on the Internet, with all the methods that we have now. The idea is knowing how far the water will go. That is pretty stressful, how far the water will rise [...] I connected cameras, so that I can see it from home [...] it is a question of getting organized. Now, we have the means, there is the *Vigicrues*[20], that's great, we can follow everything on the Internet all

---

19 The fact of being worried. See the topic of "concern" in the face of risk: [BRU 08].

20 Official French information service concerning the flood risk of the main water streams in the country (www.vigicrues.gouv.fr).

> the time, [...] I have a subscription to *Predict*[21]. When I see that the event is taking place, I see the red mass, where it is, I know where it is falling, and where to get moving or not, whether I go down there or not" [Inhabitant of Sommières, restaurant owner, about 45 years old, arrived from another region five years ago, married to a Gard native, flooded several times at her restaurant].

The two other cases we met were unanimous: their attachment to their home (the family home) bound them and it was inconceivable for them to leave. For one of these inhabitants, his departure would be forced by a poorly received eviction notice. The interview conducted with this inhabitant was extremely challenging from an emotional point of view, for he had received the eviction notice just a few days earlier. He was extremely assertive about this: otherwise he would never have left (water height in the house upstairs in 2002: four m). He disputed the decision of the administration, arguing that water did not pose real life risk.

> "There is no risk, besides. The water rises but retreats, so there is no current. This, they do not want to understand at the *Préfecture*. [...] The water rises, okay, but there is no current" [Native from Collias, about 65 years old, had returned to the commune more than five years ago, flooded in 2002].

This opposition to expert knowledge on the part of the inhabitants about such a technical subject as flood risk is a common element in many sociological studies on the perception of residents toward risk [UCR 00, COA 99]. The inhabitants regularly denounce as excessive precautionary principles on the part of the authorities in comparison to their own knowledge of local risks. For another inhabitant who is highly exposed in his home, the decision not to leave despite the repeated and serious damage in his house was exclusively motivated by the attachment he felt to this house, where he had grown up. He realized that his decision to stay here with his family despite the high risk was very costly (he clearly remembered the flood of 1988 when they had to wait for help, taking refuge on the roof of the house then covered by water). He even admitted that this generated strong tensions within his home, where his wife did not share his attachment to the place and its environment.

> "I was born here, it's true that leaving for me [...] [he lowers his voice] hurts, [...] If they told me to leave tomorrow because the house will be demolished... But I... I would get depressed! [...] Then, there is the

21 Website for issuing hydrometeorological alerts as well as overall preventive information (www.predictservices.com).

> house itself, and then there is the neighborhood that is important too, the relational aspect, the neighborhood, of course. When you have lived in a place for a while and you know almost everyone [...] If I leave this place, it would be like shooting myself in the foot, for sure. [...] Were they to offer me, I do not know, let's say, a million euros by telling me: "your house will be demolished tomorrow", but I... No, it is, it would be... Even pychologically, I would prefer to live bearing the risk of being flooded, but I, I am aware ... It's obvious. I would not go elsewhere, not at all. Then, if my wife, well, wants to get a divorce or something... But well... It is very hard for me, the idea of leaving this place" [Inhabitant of Nîmes, native, about 45 years old].

The various testimonies gathered concerning the attachment to the place and to the others, in combination with a strong will to stay despite risk encouraged us to formulate a hypothesis of possible reinforcement of attachment engendered by the flood itself. According to Bernard [BER 15]:

> "It is at the moment of their testing and their reconfiguration that attachments and supporting space–time are revealed (or constructed more radically). [...] It is at the moment of strong ruptures and reconfigurations that we able to look at them and to call them attachments".

As it was revealed in the interview excerpt above, the importance of the question of attachment to the territory is accompanied by the attachment to others, to the neighborhood, and to loved ones. In the same vein, many inhabitants of the communes of Collias and Sommières mentioned the tremendous surges of solidarity that followed the floods, suggesting that this reinforced their sense of community.

> "Paradoxically... It is quite funny because not everything is bad. There is solidarity in the street, the experience was absolutely fantastic. The neighbors we barely knew because we had been here only for a year, we became friends [...] I mean, I have pretty extraordinary memories, we picnicked in the middle of the street, on car hoods. It is funny, it creates bonds. [...] I mean, there was a solidarity that was there... which was... which was pretty extraordinary. [...] Everyone helped each other. And... As I told you at the beginning, it is the paradox of this story. For me, it is the good memories that come first". [Restaurant owner having lived in Sommières for more than ten years, around 45 years old, arrived from abroad, married to a native.]

This surge of solidarity, often repeated, went beyond political or personal splits that may be present in the communes, and created exceptions where the sense of priorities changed, and reinforced the sense of solidarity between the inhabitants.

> "There are solidarities like that, which are very important. And after there were... two which were really fabulous, first, the cleaning. Suddenly, there were a lot of people who got together, and a lot of hunters who said, "Well, let's go around all the houses, we will clean up." And there were two, three, quite smart neighbors, who said: "Well, we will not discriminate who we are angry at". Because there are unbelievable stories, there are people who cannot stand each other". [Native from Collias, returned to the commune more than five years ago.]

Attachment to a place and to others is understood according to Hennion [HEN 10, HEN 14] as "multiple links woven around, with, and by the subject on the occasion of a certain activity in which subjectivity is built in an intimate way". This appears reinforced by the experience of sharing the extraordinary flood incident. We can even go so far as to assume that this is part of the attachment that motivates people to stay in an exposed area despite the risk.

> "But the thing there is, the lesson that is to be taken here, I found, after the experience of Sommières, that... the human being becomes human again. [...] He becomes a normal person again, with a heart and a brain that work. [...] Yes, so after all [the flood] is useful for learning a couple of things, maybe. [Inhabitant of Sommières, about 45 years old, arrived from another French region more than five years ago, married to a Gard native.]

Defining moments such as flood events open a new space of possibility, develop a new space–time mosaic in the sense of Bernard [BER 15]. Attachment to the place and to others – considered not in the sense of something acquired ("attached"), but in the process of becoming ("becoming attached") – is reinforced through and by the experience of the flood, on two levels: at the individual level – in one's own body – and in the shared territory, at the collective level.

2.3.2.3. *Moving out under constraints*

The testimonies that we were able to collect from people who had given up living in a high-risk area were cases of "forced" expropriation. Cases of "voluntary" departure could not be directly addressed, what indicates one of the biases of the survey: people who had left were more difficult to meet. Nevertheless, some cases were addressed through the testimony of inhabitants concerning former neighbors who had left. But, these cases remain marginal. For example, one resident mentioned former neighbors who had left because of the flood. As far back as her memory could go, she only knew one case. These former neighbors had bought a house near the river (the Arre). Every time the river was flooded, they went to sleep at their friends' house across the street, taking care to leave everything upstairs in their home. According to the interviewee, the fear of a new flood anguished the husband, and his wife found it very tedious to have to move everything regularly for preventive reasons. These two motivations combined ushered their move to a neighboring hamlet, away from the flood zone. An inhabitant of a neighborhood in Nîmes who had suffered a dangerous incident reported several cases of neighbors who had accepted the expropriation proposal made by the commune, either because of their advanced age or – in his opinion – due to their weak attachment to the neighborhood.

> "Some people I know left, because they were old and they were fed up with having to start everything all over again... all over again from scratch, they were fed up. There are people who accepted [the expropriation proposal.] [...] There are even beautiful villas, even one-storey villas which were demolished, villas where we would have said: "But no, you had 50 centimeters in your garage, there is no reason to leave, hey." And there, the people left, after all… The experience I have here, there are some who may not have it because they arrived only later... I would say in the middle of the process, or in the meantime [Native inhabitant of a house located in the flood zone and exposed to a hazard, around 45 years old, Nîmes].

The residential trajectories observed in the Gard department show that in most cases the experience of flooding or the fact of being exposed to a flood is not a sufficient criterion in the decision to stay or leave. Overall, residential choices encompass other dimensions of life, where the question of lifestyle and organization prevails over the sole question of risk. Participant observation conducted in the field (time spent in cafes, shops, fleeting attention to local conversations) made it possible to frame the "problem" of the flood against reality, as one among others. As it was shown in other fields in flood-prone areas [DUR 14], the question of living with risk is embedded into that of a broader, more inclusive sense of living. While the issue of flooding may be a problem locally, this is not the only one. Several other subjects of concern about their living spaces favor exchanges between the inhabitants. Thus, in times of repeated field observation in public spaces[22] or overheard conversations, the question of the future of the commune in relation to urban development

22 Participant observation seeks to reveal publicly shared topics for concern by performing observation-listening sessions in social exchange places, especially in cafés. See Appendix 2 for the research methodology.

(construction of new housing, criticism of the suburban phenomenon, fear of becoming a dormitory community (or commuter town) and losing their identity, etc.) was the issue that dominated the exchanges.

2.3.2.4. *Between dramatization and relativization: from a state of vigilance to risk taking*

Wishing to remain despite the risk does not mean denying it. The people we met were well aware of the risks and most of them explained the steps they took in order to deal with them. All the people we met who had experienced a flood in their homes but still lived there had taken special measures for preventing future floods. While the traumatic experience was not dominant, admitting to a frequent anxiety, especially during the fall, was recurrent in their narratives.

> "The Gardon [...] arrived regularly in the garden. We cope with it, it is stressful, we must not say, it's not ... it is true that it's stressful, we are on alert, we do not sleep at night because we check whether it will be necessary to lift the furniture, if we have to... but I think that I have always coped with that [...]". [Collias inhabitant, around 40 years old, returned to her home commune more than five years ago].

The stories related to the fall period and seasonal stress were highly ambivalent. On the one hand, they showed an operative watchfulness [DUR 16], testifying to the existence of a local flood culture. Although the experience of the autumnal stress related to flood risk was frequently mentioned in the stories, its recurrence was often accompanied by a relativization produced by the adaptation of daily behavior.

> "The floods, yes, they do not scare me at all. Still... I would not have gone so far as to take my car when we were on red alert, but we are still watchful, we know that we must be careful. It does not appear in ten minutes... We still have time to... see it coming. [...] Then, it's useless to worry, what must happen, will happen. Otherwise, we would have to worry about the floods, we would have to worry about the terrorist threat, for example, we wouldn't live any more, we would stay hidden at home. We are still aware in these cases, but... it's raining, it's raining, we're going to stay at home, there's a blocked road, it's okay, we turn around, and that's all" [Sommières inhabitant, Gard native, arrived to the commune less than ten years ago, around 35 years old].

The story of the inhabitant from Nîmes, quoted above, who despite being exposed in his home to a strong risk and having experienced several floods (including the major flood of 1988), was so attached to his place of residence that he could not consider leaving it, is exemplary of this ambivalence between stress and relativization. On many occasions along the extensive interview, he alternated between these two poles. Thus, he offered stories of flood experiences interspersed with risk-related elements, thanks to possible adaptative countermeasures:

> "I was not worried a great deal, in fact my mother was really afraid that the house would go with the water, and as we were on the roof, that everything would fall apart. Later, we can reequip the house. Well, we are planning to do it, to put some cofferdams that make it possible... to stop the water, in fact, to prevent water from entering the, inside the house with... a system of barriers, slightly inflatable on the joints" [Nîmes resident native, about 45 years old].

The possible adaptation of behavior in times of crisis appears as a softening, counterweighting the discussion [DUR 14]. We can assume that this rhetoric device broadly expresses a resilience in the consideration of risk. The fact of relativizing by mentioning possible adaptations in the discussion helps to rationalize the speech, in the same way as making arrangements would make it possible to coexist with the possibility of a flood. The way in which each inhabitant copes with this tension (between consideration and relativization) is still highly variable. The power of relativization in the face of a situation that occurs with a certain recurrence (some native inhabitants had already experienced several important floods) can become a habit for some. Living there means having experienced several floods and having agreed to stay all the same. This idea of living in a place like this can then be likened to being able to manage the event, or to know it, sometimes even to take risks, as the following interview excerpt illustrates:

> "[Him]: Oh yes. Oh well it was very violent, it was very fast and very violent, we had only time to run, and... [...] My wife will tell you I am oblivious, because I was quiet, I started collecting my documents. [Her]: Oh yes, but he, there may be water, a storm, whatever, he does not care at all [Him]: [laughs]. [Her]: Yes, but in 2002, you nearly got caught. And even last year, that was nearly the end of you! Last year, if there had not been the girl's boyfriend, and the neighbor, he... would have been washed away. Because he did not believe that there could be so much water, and it be so strong" [Collias couple, husband is a native, wife comes from another region, about 65 years old].

If we picture risk acknowledgment on an axis, two ends can be drawn from the narratives of the inhabitants: from caution to risk taking. The speeches of the inhabitants reveal possible oscillations from one to the other, depending on the persons or sometimes on the same individual, but always depending on the situation.

Accounting for movements along this theoretical axis is still a complex challenge. In section 2.4, we will introduce one of the points of convergence between the analyses at the level of the two time scales proposed, which will help us to explain the difficulty of staying on the precautionary side permanently.

2.4. Social convergence of both scales: "the force of everyday life"

In the cross-analysis of short-term floods and long residential trajectories, the reasons for not taking into account risk, or for relativizing it, can be partly explained by what we have called "the force of everyday life". At the time of the crisis, this force is exerted by the will to continue and carry on with life as it had been organized. Numerous testimonies of inhabitants have reported this, to the point of envisioning putting themselves in danger by persisting in wanting to carry out their previously planned activities, sometimes "at all costs". The extreme endangering experience of the inhabitant mentioned above during the flood of Nîmes in 1988 can be explained by this desire to carry on with life, to continue activities according to plan.

> "I was travelling back to the institute [...] so I couldn't have missed it [...] and when I reached Nîmes, I saw that there was water, that it was raining very hard, I tried to take a street, there was water, I said to myself: "well, you are not going to ruin your car". I parked and I left on foot to reach the institute on the road to Uzes [...] Was I afraid? No, I knew that I would survive, I could not picture myself being drowned, I couldn't imagine it, the water was not so strong. On the other hand, it is true that the more I advanced on foot, at the detour of each road, street, the water level was increasing, and, why did I not turn back, I, I do not know why, I wanted to get there, I wanted to get there. [...] I do not give up easily, that's what people say about me in general... the people who know me [...] And there, same feeling, the next day, I, I, I went back, saying to myself: "You must go to the institute, you have to go to work". [Sommières inhabitant, around 45 years old, arrived from another departmental commune more than ten years ago.]

Several inhabitants who work away from their place of residence admitted that their main concern in these situations of meteorological alert (where it is strongly advised to avoid traveling as much as possible) is to be able to return home, even if they have already directly experienced the possible dangers of this action. Even when they say they adapt their behavior and are vigilant about a possible flood, the desire to return home pushes them to take to the road, sometimes in extreme conditions. The two long interview excerpts below illustrate this compelling desire to return at all costs.

> "So there was a lot of rain and I ended up driving on a small road which is often deserted, but which avoids the big arterial roads [...] and I found myself... finally... in my car, water was at the level of my door [...]. And there, frankly, I was afraid. I was scared, because it was deserted, there was... there were no cars in front, there were no cars behind, and there I said to myself "You really have to get moving". Well, I managed to get out of this place, and then, I said to myself "You have to go to a high place so that there is no water, so that's what I did. And... and so, here, I managed to climb, I absolutely wanted to go up, not to find myself blocked in the water. And there, I stayed on a parking lot while I waited it out [...] We live in... in an environment where... it is true that, as soon as it rains, now, here… we are pretty alert. Yes, no, we are used to that, we need to be organized, we must...".
>
> **Yes, you are vigilant about that, right?**
>
> "Yes, sure. Yes, Yes. [...] When I get home, that is to say if I... if we are on alert and I leave the office earlier, for example, I'm not afraid, I'm not afraid to take my car and go home. That's it, that does not scare me. What would rather annoy me is to be blocked, because when we come back home, it's already late, ... the night is near,... now… I... I do not even know what I would do if I... if I got stuck. So, maybe I would wait all night or I... well, that's just what annoys me. I do not have... I have no... when I take my car, I'm not afraid to go home, you know. I do not say to myself... that I'm going to get carried away by a wave of water, I... no".
>
> **So you say to yourself: "I am worried about getting home, about being able to go back home"?**
>
> "To arrive home, not to be stuck. Not to be worried about that. [...] But yes, the aim is to get home, ... to pick up my son and... to be all at home, actually. That's it. [Gard inhabitant, around 40 years old, arrived from another French region less than ten years ago].
>
> "On Monday morning, my mother took me to the Vocational Training Center (CFA) (something we never should have done) because it was raining and I could not take my motorcycle. In the morning there was

no problem, we took the usual road, the D792, but the water was already there, along the road on the Vigières side, but we managed to cross Remoulins, St Bonnet and all. At the CFA, classes had been canceled. I called my mother and she said, I'll tell your dad to come and pick you up. My father had to come back at noon, take the highway and pick me up, but my father... the Gardon was on the highway already. My father was stuck on the highway at the place where the Gardon goes through. It was around noon, 12.30 pm. The only thing to do was to call my uncle, who has a 4x4. My mother called him, and he managed to go through the bridge that crosses over the railway, he came from Avignon, he arrived at 3 pm. I was all alone in the meantime, my father did not have a cell phone, we could not reach him, and I was really worried. My uncle came to pick me up and on the way back, it was a disaster. The Remoulins bridge was finally blocked, they had blocked it, there was water coming from the outside, we managed to go through with the 4x4. But we crossed, anyway. All the cars had stopped, but there was already water there, when we arrived at Renault's roundabout, well, the water took us away, we started to slide with the 4x4, the water that came into the 4x4. Then, under the SNCF bridge, there was a three-meter waterfall. We stopped in the roundabout against a truck, otherwise we would have been swept away. The water started getting into the car, the electric windows were blocked, we couldn't open the windows, we broke the panes and we managed to ride on the car's roof. The helicopter came to pick us up from there, and dropped us off at the garage, where there was no water. That's when I became more worried because I saw that my father had left the car there, he had turned back to the highway, he had left the car there. I said well it's okay, my father must have walked back. The helicopters had made a field that blocked cell phones, so as not to interfere with their radios. My cell phone did not have enough signal, I had it in my mouth and I was swimming, then I climbed on the roof of the 4x4. After that, we went in the direction of the highway bridge toward Avignon so as to have cell phone signal, we must have walked about a kilometer on foot, I thought maybe they could come pick us up by another road. I called my mother in Saint Hilaire, she told me that we were stuck anyway, we had to walk back. We went through the D192, but there, it was a disaster, at some places there were two meters of water, we had to swim, with the current, the water went over the road, but that was the rainwater coming down from the mountains. This was where there was the overturned car the previous day. We swam, I couldn't touch the ground with my feet, it was unbelievable, I clung with the feet to the vines so as to be able to detach myself. At some places there was no water, at some other places water was extremely high. The hardest thing was there. Then, we went back quietly to Saint Hilaire. We wanted to go there because we could not go elsewhere, there we knew the road and we could go through the fields, but we did not know how things would turn out. My father, yes, from there, he cut short directly through the fields. I wanted to go home, I wanted to see my father, anyway I did not know what had happened to him. When I reached my mother, she told me that my father had come home, yes, but where would I sleep? I preferred to go home. We both said, we will go back, there was nothing else to be done. No matter how, we had to go home" [Inhabitant of Saint Hilaire, 17 years old, apprentice].

The recurrence of meteorological alerts in the department helps to explain this risk taking, which is often taken for granted in the face of a flood. It is hard to upset a schedule every time a weather alert is given. Apart from professional obligations, widely spread doubts about the accuracy of such alerts helps to account for the forms of distrust expressed by many inhabitants. First of all, zoning is widely disputed. Meteorological alerts are given at the departmental level, when in most cases floods only apply to a very small part of it. The multiplication of situations of vigilance associated with the fact that their implementation has been recent[23] leads many inhabitants to think that it is a kind of abusive precaution on the part of the authorities in order to counter possible complaints against any possible deficiencies in anticipating the event.

This *force of everyday life* identified in the short term of the event can be extended to the long term of the inhabitant trajectories. As discussed above, numerous other criteria prevail in the acceptance of living in a flood-prone area. Among these criteria, the question of lifestyle choice and living environment, as well as the question of the attachment to places and to the others make part of the everyday life of inhabitants. These elements justify the fact of appreciating living there and thus continuing to do so. The living environment, the lifestyle habits and the attachments that bind them to their environment and to others constitute the daily life of the inhabitants. Even if certainty about new future floods is mostly admitted, confronted with what is perceived as exceptional, these constitutive elements of everyday life are stronger than the risk of having to decide whether to stay or leave. It is this daily routine that occupies the larger part of the inhabitants' time. The flood remains an exceptional occurrence and cannot always interrupt everything that constitutes everyday life. When referring to their flooding experiences, people frequently report that they would never have imagined that such a flood would be possible, but still describe the floods they may have known mainly as exceptions, implying that there certainly will not be new ones of such magnitude before long.

23 Device implemented since October 2001 (http://vigilance.meteofrance.com).

> "1994 ... there was water. In 2002, in 1982, there was water, or in 1983, there had been water. But never like that [in reference to the episode of autumn 2014]. It was exceptional" [Native resident from Vigan, back in the commune for more than ten years].

For some, even those who had been flooded in their homes, this relativization of the importance of the issue with regard to its rarity assumes the form of a probabilistic calculation.

> "But, between 1958 and 2002, that is still 44 years, well, if we experience a big flood every 44 years... that makes a little helicopter ride every 44 years" [Inhabitant of Sommières, about 65 years old, native having returned to the commune ten years ago, sought refuge on his roof and was rescued by helicopter with his family in 2002].

Stating that a helicopter tour every 44 years is acceptable means assessing the 44 years when it was possible to live there without floods by taking advantage of the elements appreciated in the fact of living there, part of everyday life. In the narratives of locals valuing their place of residence, the flood remains confined to a register of the exceptional against all the other criteria of living there on a daily basis, which are more defining aspects.

2.5. Conclusion

Over the long term of residential trajectories, the collected narratives of the inhabitants question the primacy of a rational logic in favor of a logic of attachment to the place and to the people who live there. What predominates in the explanation of residential choices of the inhabitants from Gard are the following criteria: the appreciation of the living environment and the lifestyle associated with it, and a sense of belonging to a community [HAN 83]. The attachment to the place and to people reinforces, and even explains, these criteria. This attachment is what has the strongest explanatory power for people's willingness to stay despite the risk for those whose homes are particularly exposed or have already been impacted. The solidarities that the event stimulates help to reinforce this attachment to the place and those with whom it is shared. In residential choices, these criteria overlook the issue of flooding, whose severity is still perceived as rare. The exceptional character of the flood is not strong enough in the face of the criteria that prevail in the long term of inhabiting a place. In the short term, the narratives of people who experienced a flash flood reveal the diversity of experiences and the difficulty in understanding the complexity of situations, which all require a specific response to very local conditions. The reaction speed of small watersheds imposes specific spatial and temporal constraints such that a generic flood experience is not able to be anticipated. Finally, the daily schedule appears both as a discouragement to paying attention (attention which should be focused on the observation of environmental signs), and as a constraint, particularly related to professional activities. While certain factors – such as responsibility for children – encourage precaution, a willingness to pursue the usual activities, for instance, sticking to the normal course of events, stirs up a force that hinders the assessment of the real danger of the situation. Thus, the *force of everyday life* appears as a form of convergence between the two time scales we have explored, which may be used to evaluate the social effects of floods on mobility.

2.6. Bibliography

[AUT 10] AUTHIER J.-Y., BONVALET C., LEVY J.-P. (eds), *Élire Domicile. La construction sociale des choix résidentiels*, Presses universitaires de Lyon, Lyon, 2010.

[BEC 63] BECKER H., *Outsiders*, The Free Press, New York, 1963.

[BEL 68] BELL W.L., "The city, suburb and a theory of social choice", in GREER S. *et al., The New Urbanization*, Saint Martin's Press, New York, 1968.

[BER 97] BERTAUX D., *Le récit de vie*, Armand Colin, Paris, 1997.

[BER 15] BERNARD G., "Ruptures urbaines: une pragmatique spatio-temporelle", *Parcours anthropologiques*, Centre de recherches et d'études anthropologiques, pp. 46–64, 2015.

[BES 10] BESSIN M., BIDART C., GROSSETTI M. (eds), *Bifurcations. Les sciences sociales face aux ruptures et à l'événement*, La Découverte, Paris, 2010.

[BON 99] BONVALET C., GOTMAN A., GRAFMEYER Y. (eds), *La famille et ses proches. L'aménagement des territoires*, Presses universitaires de France, Paris, 1999.

[BRU 94] BRUN J., FAGNANI J., "Lifestyles and locational choices–trade-offs and compromises: a case-study of middle-class couples living in the Ile-de-France Region", *Urban Studies*, vol. 31, no. 6, pp. 921–934, 1994.

[BRU 08] BRUNET P., "De l'usage raisonné de la notion de 'concernement': mobilisations locales à propos de l'industrie nucléaire", *Natures Sciences Sociétés*, vol. 16, no. 4, pp. 317–325, 2008.

[CHA 07] CHARDONNEL S., "Time-geography: individuals in time and space", in SANDERS L. (ed.), *Models in Spatial Analysis,* ISTE Ltd, London and John Wiley & Sons, New York, 2007.

[CHA 13] CHARDONNEL S., DUVILLARD S., SGARD A. *et al.*, "Devenir propriétaire loin des métropoles: entre contraintes de mobilité et choix de vie", in GERBER P., CARPENTIER S. (eds), *Mobilités et modes de vie. Vers une recomposition de l'habiter*, Presses universitaires de Rennes, Rennes, 2013.

[COA 99] COANUS T., DUCHENE F., "Les relations des gestionnaires du risque urbain avec les populations riveraines. Critique d'une certaine idée de la communication", *Annales des Mines, Responsabilité et Environnement*, no. 13, pp. 5–17, 1999.

[COL 98] COLLOMB P., *Les français et l'environnement*, Presses Universitaires de France, Paris, 1998.

[CRE 13] CREUTIN J.-D., BORGA M., GRUNTFEST, E. *et al.*, "A space and time framework for analyzing human anticipation of flash floods", *Journal of Hydrology*, no. 482, pp. 14–24, 2013.

[DE 03] DE VANSSAY B., COLBEAU-JUSTIN L., "Méthodologies de retour d'expérience post-catastrophe. Catastrophes naturelles et technologiques", in MOSER G., WEISS K. (eds), *Espace de vie. Aspect de la relation homme-environnement*, Armand Colin, Paris, 2003.

[DOB 95] DOBRE M., *L'opinion publique et l'environnement*, IFEN, Paris, 1995.

[DUB 06] DUBOIS N., MOCH A., "L'automobile et le sentiment d'être chez-soi", *Revue Européenne de Psychologie Appliquée/European Review of Applied Psychology*, vol. 56, no. 2, pp. 95–107, 2006.

[DUC 00] DUCHENE F., MOREL-JOURNEL C., "Riverains de cours d'eau et gestionnaires du risque, un dialogue impossible?", *Géocarrefour*, vol. 75, no. 3, pp. 221–226, 2000.

[DUR 14] DURAND S., Vivre avec la possibilité d'une inondation? Ethnographie de l'habiter en milieu exposé... et prisé, PhD Thesis, Université Aix-Marseille, 2014.

[DUR 16] DURAND S., RICHARD-FERROUDJI A., "Du principe de prévention aux cultures locales du risque inondation", in RUDOLF F. (ed.), *Les Villes à la croisée des stratégies globales et locales des enjeux climatiques,* Presses Universitaires de Laval, Quebec, 2016.

[FAB 03] FABUREL G., "Lorsque les territoires locaux entrent dans l'arène publique. Retour d'expériences en matière de conflits aéroportuaires", *Espaces et Sociétés*, vol. 115, pp. 123–146, 2003.

[GRA 10] GRAFMEYER Y., "Approches sociologiques des choix résidentiels", in AUTHIER J.Y., BONVALET C., LEVY J.-P. (eds), *Élire domicile. La construction sociale des choix résidentiels*, Presses Universitaires de Lyon, Lyon, 2010.

[GRO 04] GROSSETTI M., *Sociologie de l'imprévisible. Dynamiques de l'activité et des formes sociales*, Presses Universitaire de France, Paris, 2004.

[GRO 06] GROSSETTI M., "L'imprévisibilité dans les parcours sociaux", *Cahiers Internationaux de Sociologie*, vol. 210, pp. 5–28, 2006.

[GRO 16] GROSSETTI M., "L'imprévisibilité dans le monde social", in LEVY S. (ed.), *Complexité et désordre. Éléments de réflexion*, Presses Universitaires de Grenoble, Grenoble, 2016.

[GUE 06] GUEYMARD S., "Facteurs environnementaux de proximité et choix résidentiels", *Développement durable et territoires,* dossier 7, 2006.

[HAG 70] HAGERSTRAND T., "What about people in regional science?", *Regional Science*, vol. 24, pp. 7–24, 1970.

[HAN 83] HANNERZ U., *Explorer la ville,* Les éditions de minuit, Paris, 1983.

[HEN 10] HENNION A., "Vous avez dit attachements?", in AKRICH M., BARTHE Y., MUNIESA F. *et al.* (eds), *Mélanges en l'honneur de Michel Callon*, Presses des Mines, Paris, 2010.

[HEN 18] HENNION A., D'une sociologie de la médiation à une pragmatique des attachements : retour sur un parcours sociologique au sein du CSI, *SociologieS, Théories et recherches*, online 25 June 2013, accessed 3 May 2018.

[HUG 96] HUGHES E.C., "Carrières, cycles et tournants de l'existence", *Le regard sociologique*, EHESS, Paris, 1996.

[KAU 96] KAUFMAN J.C., *L'entretien compréhensif,* Nathan, Paris, 1996.

[LIN 03] LINDELL M.K., PERRY R.W., *Communicating environmental risk in multiethnic communities*, Sage Publications, Thousand Oaks, 2003.

[LIN 12] LINDELL M.K., PERRY R.W., "The protective action decision model: theoretical modifications and additional evidence", *Risk Analysis*, vol. 32, no. 4, pp. 616–632, 2012.

[MAR 99] MARESCA B., HEBEL P., *L'environnement: ce qu'en disent les français,* La Documentation française/Crédoc, Paris, 1999.

[MIL 95] MILETI D.S., "Factors related to flood warning response", US–Italy Research Workshop on the Hydrometeorology, Impacts, and Management of Extreme Floods, report, pp. 1–17, 1995.

[NEI 96] NEISSER U., "Remembering the earthquake: direct experience vs. hearing the news", *Memory*, vol. 4, no. 4, pp. 7–358, 1996.

[NOR 92] NORRIS F.H., KANIASTY K., "Reliability of delayed self-reports in disaster research", *Journal of Traumatic Stress*, vol. 5, no. 4, pp. 575–588, 1992.

[PIN 02] PINKSTER F., VAN KEMPEN R., *Leefstijlen en Woonmilieuvoorkeuren*, Urban and Regional Research Center, University of Utrecht, Utrecht, 2002.

[RUI 07] RUIN I., Conduite à contre-courant. Les pratiques de mobilité dans le Gard: facteur de vulnérabilité aux crues rapides, PhD Thesis, Université Joseph Fourier/Grenoble I, 2007.

[RUI 08] RUIN I., CREUTIN J.-D., ANQUETIN S. *et al.*, "Human exposure to flash floods–relation between flood parameters and human vulnerability during a storm of September 2002 in Southern France", *Journal of Hydrology*, vol. 361, no. 1, pp. 199–213, 2008.

[RUI 14] RUIN I., LUTOFF C., BOUDEVILLAIN B. *et al.*, "Social and hydrological responses to extreme precipitations: an interdisciplinary strategy for postflood investigation", *Weather, Climate and Society,* vol. 6, no. 1, pp. 135–153, 2014.

[THO 09] THOMAS M.P., PATTARONI L., KAUFMANN V., "Modes de vie et mobilité quotidienne : quelles relations avec les choix résidentiels ?", *Colloque MFFS, Les interactions entre mobilités quotidienne et résidentielle à l'épreuve des nouvelles pratiques sociales*, Luxembourg, March, 26–28, 2009.

[VAN 05] VAN GIEZEN A.E., ARENSMAN E., SPINHOVEN P. *et al.*, "Consistency of memory for emotionally arousing events: a review of prospective and experimental studies", *Psychology Review*, vol. 25, no. 7, pp. 935–953, 2005.

[WEI 89] WEINSTEIN N.D., "Effects of personal experience on self-protective behavior", *Psychological Bulletin*, vol. 105, no. 1, pp. 31–50, 1989.

[WYN 92] WYNNE B., "Uncertainty and environmental learning: reconceiving science and policy in the preventive paradigm", *Global Environmental Change,* vol. 2, pp. 111–127, 1992.

3

The Challenges of Flash Flood Forecasting

3.1. The spatial and temporal scales of flash floods

The consequences of flash floods in Mediterranean areas during intense rainfall periods can be dramatic, as shown by the episodes that affected the Gard region in 2002 [DEL 05, RUI 08], the Var in June 2010 [RUI 14], the south of France in 2014 or the Alpes-Maritimes in October 2015. Preventing flash floods in time (as soon as possible) and space (as accurately as possible) is a powerful means for reducing the economic and human consequences of these events.

Before dealing with the question on how to prevent these devastating events, it is important to define what a flash flood is. A first definition drawn from Gaume *et al.* [GAU 09], quoted from an IAHS-UNESCO-WMO report (1974), describes it as: "a fast flow, with a high peak discharge value, produced by intense rainfall, generally limited in their spatial extent". In a study presenting a compilation of European flash floods, Gaume *et al.* [GAU 09] indicated that "... extreme floods produced by severe and stationary rainfall can be considered as flash floods". They pointed out that flash floods are often associated with intense rainfall exceeding 100 mm in a few hours, and affecting a limited area. Nonetheless, these same authors observed that the rainfall generating these floods can last for a long time (approximately 24 h) and be of moderate intensity, leading to the accumulation of several hundreds of millimeters of rainfall. Gaume *et al.* also showed that flash flood specific peak discharges range from 0.5 to 40 m^3 s^{-1} km^{-2} [GAU 09]. In order to clarify the resonance between rainfall forcing and the hydrological responses of affected catchments, we will stick to the definition of flash floods suggested by Braud *et al.*

Chapter written by Isabelle BRAUD, Béatrice VINCENDON, Sandrine ANQUETIN, Véronique DUCROCQ and Jean-Dominique CREUTIN.

[BRA 14]: the length of time lapse before peak discharge of the hydrograph[1] is reached (the duration between the beginning of a significant increase in discharge and the peak value) must be of a few hours for watersheds of 1–100 km^2 and lower than 24 h for watersheds of about 1,000 km^2. The hydrological responses of watersheds should also present sufficient specific discharge values[2], which we will set at 0.5 m^3 s^{-1} km^{-2} (see an example in Figure 3.3(c)).

Inspired by Delrieu *et al.*'s research [DEL 05], for each of the different studied watersheds, Figure 3.1 introduces the characteristic spatial and temporal scales for rainfall and associated precipitating systems leading to floods. Thunderstorms and mesoscale convective systems (MCS) are at the source of flash floods and, respectively, affect watersheds smaller than 100 km^2 and 1,000 km^2, with very short time to peak (up to a few hours for watersheds of a size <100 km^2), which leaves little time for issuing an alert or for implementing evacuation and protection measures for the population. On the other hand, studies and compilations have shown that specific peak discharges are generally higher on smaller watersheds than on larger ones (Figure 3.2), which reinforces the danger of flash floods in smaller watersheds.

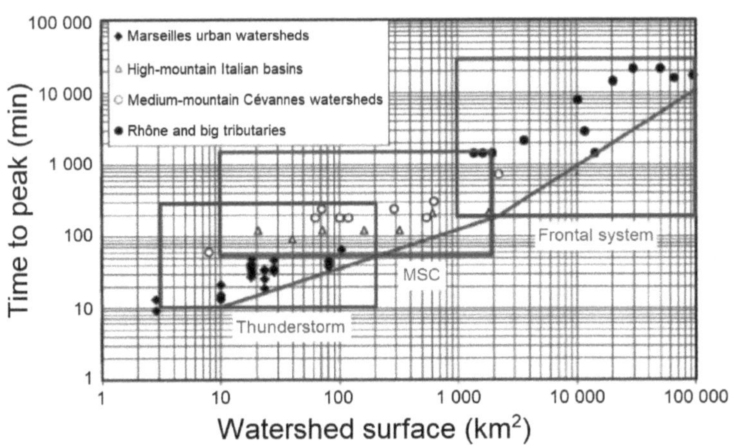

Figure 3.1. *Characteristic scales representing the response time of the Mediterranean and Alpine basins, during peak rising (time lapse between the rainfall's gravity center and the peak discharge), as a function of the basin's size. The red line outlines the envelope curve of the lowest peak times. Blue rectangles gather the characteristic scales of different types of precipitation systems: thunderstorms, mesoscale convective systems (MCS) and frontal systems (figure adapted from [DEL 16]). For a color version of this figure, see www.iste.co.uk/lutoff/hydrometeo1.zip*

1 Dicharge time evolution at a given station measured along a specific time period (for example, flood hydrograph).

2 Discharge divided by the area of the catchment for which discharge is measured.

Therefore, typical flash flood scales usually cover a few km², and peak times may oscillate between a few minutes and a few hours. Nevertheless, some flash floods can affect relatively large territories. In fact, during the Gard episode of September, 2002, the area affected by rainfall accumulations of more than 200 mm in 24 h was about 5,500 km² [DEL 05]. As we will see in the rest of the chapter, this renders the monitoring and observation of this type of event rather difficult.

Creutin and Borga [CRE 03] pointed out that conventional networks for measuring rainfall (pluviographs) and discharges (hydrometric stations) are not able to monitor hydrometeorological events leading to flash floods. In fact, the size of intense rainfall convective cells is often smaller than the interdistance between pluviographs. Besides, in general, the operational network of hydrometric stations measures watersheds of a size greater than 50–100 km² (black dots in Figure 3.2). Creutin and Borga [CRE 03] highlighted the contribution of meteorological radars to the study, the understanding and the prediction of flash floods. These radars can indeed provide rainfall fields[3] (Figure 3.3(a)) with a spatial resolution of about 1 km² and a time resolution of a few minutes, which are compatible with typical flash flood spatial and temporal scales.

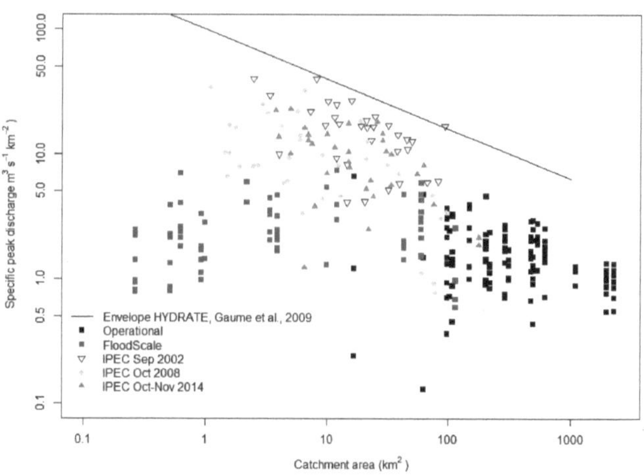

Figure 3.2. *Specific peak discharge depending on the watershed size for the Cévennes region. Blue and green triangles, as well as blue dots, correspond to postevent survey values collected for the Cévennes events of September 2002, October 2008 and October 2014. Black dots correspond to the values obtained from the discharge measurement operational network (Banque Hydro) and the red dots to a network of research stations (FloodScale project [BRA 14]). The continuous line is the envelope curve suggested by Gaume et al. [GAU 09] (figure adapted from [BRA 16]). For a color version of this figure, see www.iste.co.uk/lutoff/hydrometeo1.zip*

3 Map of rainfall spatial distribution.

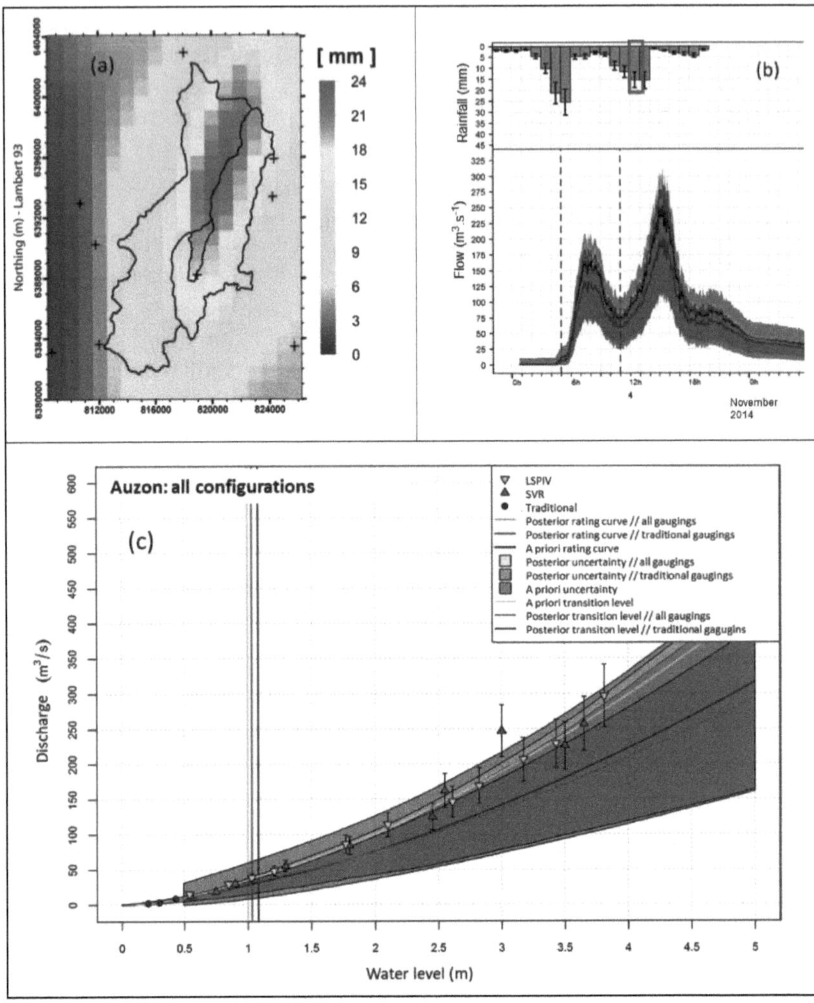

Figure 3.3. *a) Example of an estimated rain field (Ardèche on November 11, 2014 at 1–2 pm), obtained through the merging of radar observations and rain gauges, and the corresponding estimation error for the Auzon basin (116 km^2) (see error bar in the red rectangle of Figure 3.3(c)); b) error propagation from the rating curve to the flood hydrograph of November 11, 2014 at Auzon's Flood Scale station. Blue, green and pink stripes correspond to errors associated with discharge estimation as a function of gauging used for the estimation of the rating curve (blue: only a priori knowledge about the section's hydraulics; green: a priori hydraulics and traditional gauging only; pink: all gauging, including mobile radar flood gaugings and gauging obtained at the LSPIV station (image analysis) (figure adapted from [BRA 16]); c) rating curve (water level–discharge relationship). For a color version of this figure, see www.iste.co.uk/lutoff/hydrometeo1.zip*

Various studies have contributed to the emergence of methods for merging radar observations with pluviographs, so as to improve the accuracy of rainfall field quantitative estimation [TAB 07a, TAB 07b, LED 14, BUD 16]. See example in Figure 3.3(a).

In a hydrometric station, discharges are calculated from the continuous measurement of the river's water level. Specific discharge measurements, known as gaugings, are required to establish a relationship between the water level and the discharge. This is called a rating curve, and its purpose is to translate water levels into discharges (Figure 3.3(b)). Traditional discharge measurements are intrusive, in that they require the insertion of a device (current meter, ADCP[4], etc.) in the water and/or require the presence of staff in or near the river. For the safety of the staff and the equipment, it is impossible to carry out these measurements during the flood. Therefore, rating curves are often extrapolated for high flows, far beyond the range of gauged values (Figure 3.3(b), blue or green curve), which leads to very high uncertainties regarding flood discharges (Figures 3.3(b) and (c)). Recent advances have, nonetheless, led to the development of non-contact methods for measuring flood discharges, using mobile radars (Surface Velocity Radar [SVRs] [WEL 16], Large Scale–Particle Image Velocimetry [LSPIV] stations) equipped with cameras that film the watercourse and make it possible to infer water surface velocities by means of image analysis, and if we know the measurement section, we can then deduce the discharges [HAU 08, LEC 10]. Figures 3.3(b) and (c) illustrate the contribution of this type of measurement, which favors a better estimation of flooding discharges because of the reduction of uncertainty.

But flash floods often take place in very small, frequently ungauged watersheds, which do not have a hydrometric station. Therefore, other techniques are required to estimate peak discharges. One method aims at estimating the peak discharge by means of postevent field surveys. The methodology exploits flood marks[5] to estimate the maximum water level. Field surveys are conducted to collect data not only about the river cross-section[6], but also about the slope of the river water line[7], the roughness of the river bed, allowing to derive estimates of the peak discharge [GAU 08] (see the blue and green dots in Figure 3.2). When available, amateur videos collected on the web can also be exploited by means of the LSPIV technique so as to estimate flood discharges [LEB 16b, LEC 16]. Interviews with witnesses complete this geomorphological data providing information concerning the temporality of the phenomenon, that is to say, the estimate of the flood peak time. These interviews, developed with our social science colleagues were conducted

4 Acoustic Doppler Current Profiler.

5 Marks left by the flood (for example vegetation debris deposited on trees).

6 Waterstream geometry in a perpendicular orientation with regard to the stream's direction.

7 The water line slope is the river's surface slope in the longitudinal sense.

together with them in the field, and have become essential for better understanding the reactions of the population, as well as their adaptation to critical situations (see Chapter 2 and Ruin *et al.* [RUI 14] for the episode in the Var region, 2010).

The characterization and ranking of hydrometeorological processes leading to flash floods has been at the heart of several collaborative projects at national, European and international levels (FLOODSite, HYDRATE[8] [BOR 12] projects); HyMeX[9] program [DRO 14]; FloodScale project[10] [BRA 14, BRA 16]). All of these have sought to refine the knowledge of the hydrological processes involved in flash flood events and to develop forecasting chains, so as to anticipate the protection of populations. As discussed above, the spatiotemporal variability of rainfall is the first factor that is important to document at the characteristic scales of flash floods, but other factors that have proved important to explain the magnitude of the hydrological response [BOR 10] also include surface conditions, such as topography, initial soil moisture and land use.

Despite these observation and comprehension difficulties, flash flood forecasting systems were introduced as early as the 1970s, and have been perfected over the past decades. Before providing an overview of the available forecasting methods, we will first describe prediction systems and the barriers associated with flash flood forecasting.

3.2. How can we anticipate flash floods?

This section introduces the different stages of flood forecasting, followed by the specificities and singular difficulties related to flash flood forecasting.

3.2.1. *Flood forecasting chain elements*

A flood forecasting chain at the outlet of a watershed contains several elements as follows (Figures 3.4–3.6):

– a module that provides an estimate of observed and future rainfall (if future rainfall estimates are not available, we generally assume that future rainfall will be equal to zero) (Figure 3.4(a));

8 Hydrometeorological data resources and technologies for effective flash flood forecasting, http://www.hydrate.tesaf.unipd.it/index.asp?sezione=Home.

9 Hydrological Cycle in the Mediterranean experiment, https://www.hymex.org/index.php?lang=english&page=workshops.

10 Multiscale observation and modeling for flash flood comprehension and simulation, https://floodscale.irstea.fr/.

– a hydrological model that transforms the rainfall input into discharge at the outlet of the catchment. This model is a more or less simplified representation of the hydrological processes that enable the transformation of rainfall into discharge. The model includes a number of parameters that should be estimated. These parameters are generally fitted so that the model reproduces a time series of existing discharge data in the best possible way. This is known as model calibration (Figure 3.4(b));

– in a real-time application, the hydrologic model is fed by previous and forecasted rainfall. It also produces discharge simulations for a lead time *L* (Figure 3.5). These simulations can be regularly updated, thus making it possible to refine the forecast (Figure 3.6);

– these simulations are analyzed and the flood forecasting service determines whether it is necessary to issue an alert, for instance, if the forecasts contemplate an exceedance of predefined warning thresholds (Figure 3.4(c)). These thresholds can either be determined using a model climatology and the estimation of flood quantiles (e.g. [JAV 16]), or directly defined by the watershed's managers.

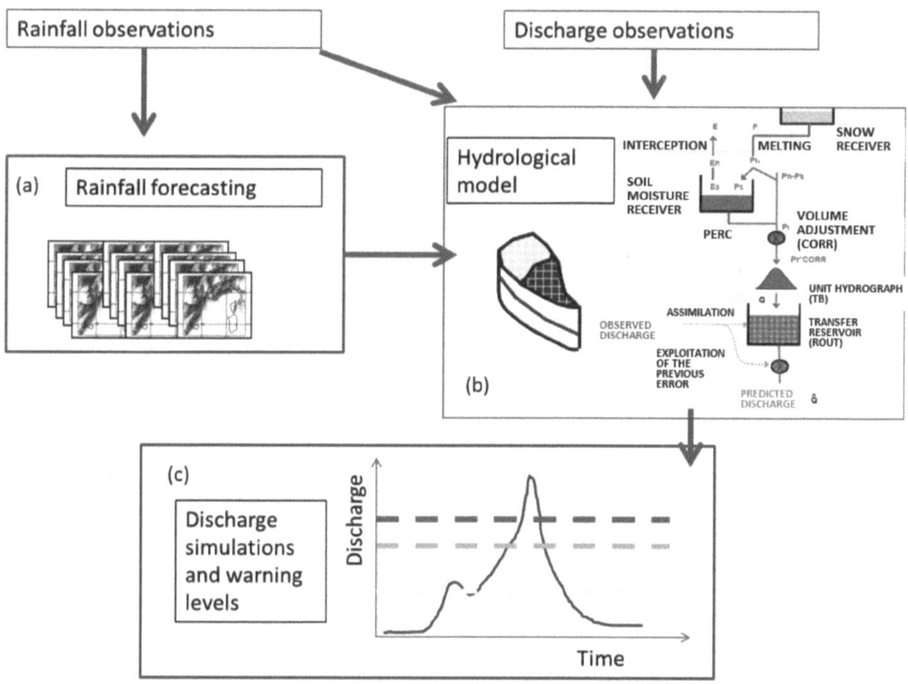

Figure 3.4. *Diagram of the principle of a flood forecasting system. For a color version of this figure, see www.iste.co.uk/lutoff/hydrometeo1.zip*

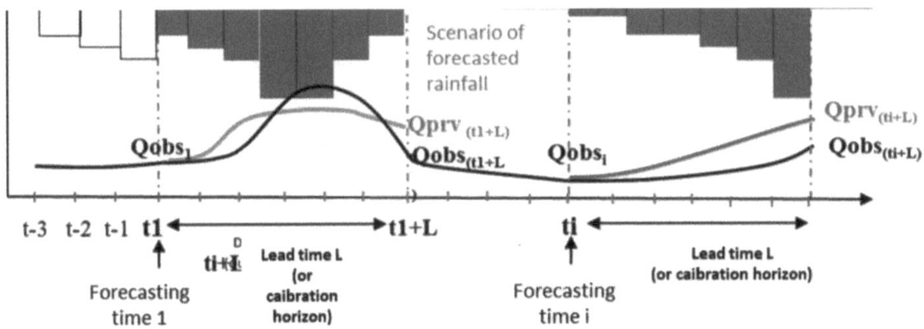

Figure 3.5. *Illustration of the notion of lead time L; at time ti, a discharge forecast is calculated for the ti+L lead time, which is the forecast's horizon. The hydrological model is set calibrated using discharge values calculated at ti+L (figure adapted from [AHR 17])*

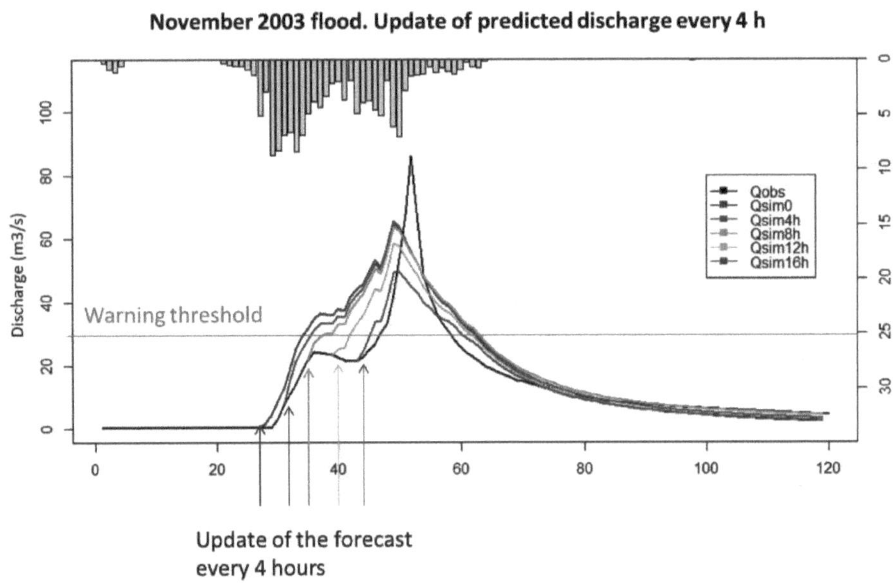

Figure 3.6. *Illustration of the update of the forecast with a regular time step. The figure shows the observed flow (Qobs) and the one simulated at 0 h forecast lead time (Qsim0), then updated every 4 h (Qsim4h .... Qsim16h) (figure adapted from [AHR 17]). For a color version of this figure, see www.iste.co.uk/lutoff/hydrometeo1.zip*

At European level, the European Flood Awareness System (EFAS) was developed by the JRC-ISPRA, under the authority of the European Commission. Lauched after the floods of the Elbe and the Danube in 2002, the aim was to have a common tool at European level so as to improve flood anticipation, particularly for cross-border rivers and thus circumvent the heterogeneity of national flood forecasting tools. This hydrometeorological forecast chain is based on LISFLOOD [THI 08], a hydrological model working with weather forecasts from the European Center and the German Meteorological Center (DWD). The study conducted on the September 2002 event in the Gard region [YOU 08] (Figure 3.7) showed that the map of threshold exceedance forecasted for September 7, 2002 at noon for the following 36 h highlighted several areas in the Gard forecast with a maximum level of risk.

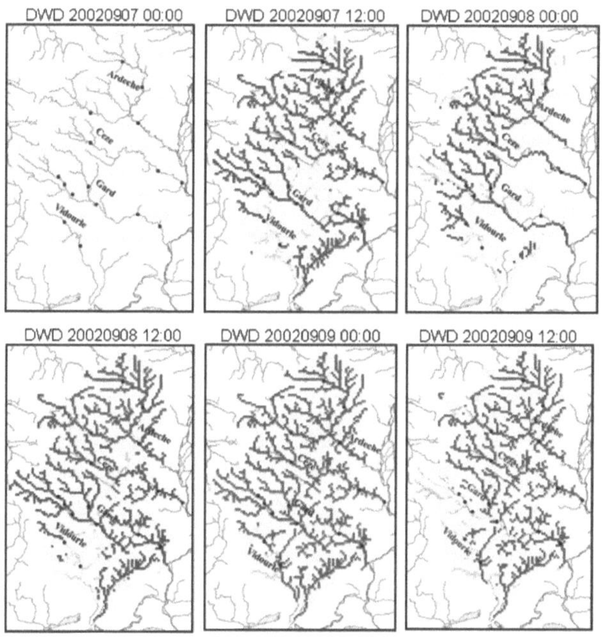

Figure 3.7. *Severity threshold (color) simulated with EFAS hydrometeorological chain. This figure shows how the Gard flood could have been anticipated in September 2002 by using this chain. Every box describes the forecasted warning level 12 h after the indicated time. This figure clearly shows that after September 7, at 12 pm, the model anticipated a severe or important hydrological situation (we must bear in mind that the flood's peak occurred between 10 pm on September 8, and 8 am September 9, for a total of cumulated rainfall of 640 mm in 24 h). The colors correspond to the following watch levels: green: water level higher than normal, but no flood expected; yellow: high water level, but no flood expected; red: strong probability of water levels reaching bankfull discharge; purple: strong probability of having overflows, even a severe flood (figure adapted from [YOU 08]). For a color version of this figure, see www.iste.co.uk/lutoff/hydrometeo1.zip*

Since 2004, flood forecasting in France has been managed by the SCHAPI (*Service Central d'Hydrométéorologie et d'Appui à la Prévision des crues*, French Central Hydrometeorology and Flood Forecasting Support Service) and the SPC (*Services de Prévision des Crues*, Flood Forecasting Services) network [VIN 07]. Rainfall forecasting is the domain of *Météo-France*. Flow forecasts are issued on so-called "regulatory" river reaches, that is to say those on which a forecast is made via the *VigieCrues* Website (Figure 3.8). These regulatory sections cover watersheds larger than 50–100 km^2.

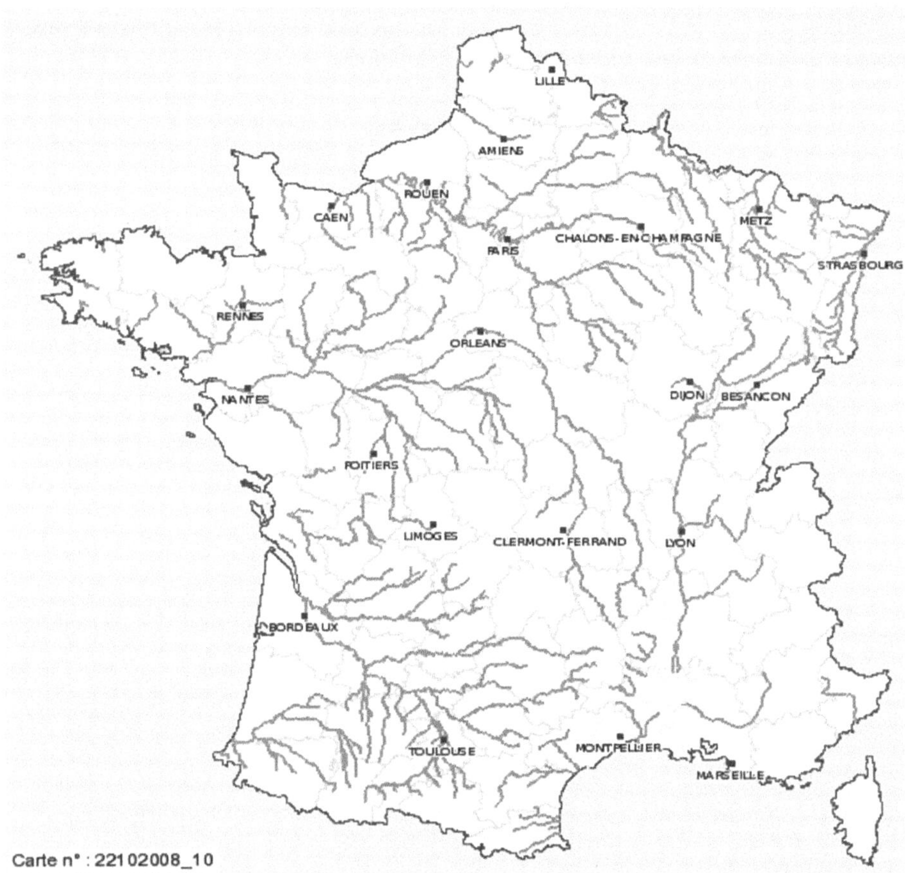

Figure 3.8. *Example of a map published on the Vigie Crues site, concerning a case study for the MobiClimEx project. Map published on October 22, 2008. For a color version of this figure, see www.iste.co.uk/lutoff/hydrometeo1.zip*

3.2.2. *Difficulties of flash flood forecasting*

With regard to the anticipation of flash floods, this approach faces several difficulties:

– spatial and temporal scales that are necessary for the prediction of flash floods are far more detailed than those used in "standard" forecasting, as described above;

– given the small size of watersheds where flash floods occur, the accuracy level required for rain weather forecasts is greater than for larger watersheds. The location of intense rainfall cells, their intensity and their timeframe must be as precise as possible. A geolocation error of about 10 km might imply that the expected flood ultimately affects a neighboring watershed;

– since flash floods mostly occur in small ungauged watersheds, the lack of discharge observations hinders the implementation and calibration of the hydrologic models previously described;

– considering the quick response time of the basins, anticipation and protection times are very short, which may limit the practical utility of the forecast.

In order to overcome these difficulties, distributed hydrological models (for instance, describing basins in detail) are necessary. These models often feature more parameters than the models used in "standard" forecasting. The parameter estimation is made difficult as it is not possible to calibrate the model at the scales of interest because data are insufficient. Therefore, it is necessary to implement calibration strategies on the gauged basins and then to extrapolate these parameters for ungauged watersheds by means of a regionalization approach. Also, some other models without calibration are used. Since rainfall intensity is often high and discharges large, it might be tempting to oversimplify models, assuming that all the rainfall is transformed into runoff. We would therefore be faced with a high rate of false alarms (alert triggering despite the fact that there is no immediate danger), which would eventually discredit the forecast and inspire a lack of trust on behalf of the users. On the other hand, out of 60 watersheds exposed to flash floods, Marchi *et al.* [MAR 09] showed that the average runoff coefficient (ratio between the volume that flowed to a watershed during the flood and the amount of rainfall received by this watershed) is barely 35 ± 18%, which shows that the watersheds may have a high rain holding capacity and that it is not all the rain that contributes to the flood.

3.3. Cutting-edge tendencies in flash flood forecasting

This section will introduce the state of the art on the prediction of intense rain events, then of flash floods, and will end up with a focus on the flash flood systems deployed in France.

3.3.1. *The forecasting of intense rainfall events*

As indicated in section 3.1, storm systems are meteorological phenomena leading to accumulations of significant rainfall during a brief time lapse, thus generating flash floods. These storm systems are particularly difficult to forecast due to the nature of the spatial and temporal scales involved: the storm cells have a lifetime of about half an hour and a spatial extension of a few kilometers. The extrapolation methods of radar observations, based on the monitoring of observed storm cells and the extrapolation of their movement, become inefficient after a half hour period.

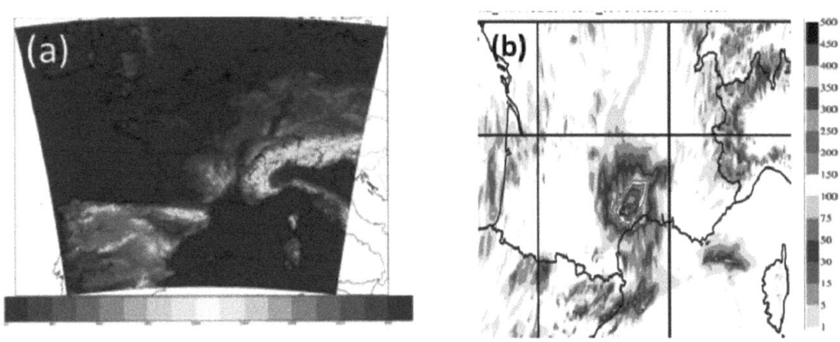

Figure 3.9. *a) Domain of the AROME model in its operational configuration for France, resolution set at 1.3 km. b) Example of AROME forecast, describing rain cumulation (in mm) for a 24-h period from September 17, 2014 at 00 UTC to September 18, 2014 at 00 UTC (analysis on September 17, 2014 at 00 UTC). For a color version of this figure, see www.iste.co.uk/lutoff/hydrometeo1.zip*

Predicting these events involves using Numerical Weather Prediction (NWP) models, which describe the evolution of atmospheric variables (pressure, wind, temperature and humidity). The most detailed meteorological models are currently attaining a kilometric horizontal resolution, which enables them to better describe the terrain and to represent the physical processes at stake. Their performance in forecasting storm events can therefore be significantly improved, particularly if they benefit from detailed initial conditions. This is the case for the AROME model [SEI 11; BRO 16] implemented by Météo-France. Its horizontal resolution has been of 1.3 km since April 2015, for an area covering a large part of Western Europe (Figure 3.9(a)). The representation of cloud processes and storm cells in terms of size, number, intensity and life cycle is more realistic and overall leads to a better forecast of rainfall totals (Figure 3.9(b)).

A key factor for the performance of these models is an accurate knowledge of the initial state of the atmosphere. A good description of the initial wind and moisture fields in the first kilometers of the troposphere[11] is particularly necessary for a realistic simulation of the intensity and location of precipitating systems [DUC 02]. Meteorological observations are used for updating the initial conditions of high temporal frequency models. So-called "data assimilation" methods make it possible to readjust the scenario resulting from a short-term numerical forecast provided with a regular frequency. Broadly different types of observations can be indirectly related to variables that were simulated by the meteorological model. For example, data measured by meteorological radars (Doppler wind and reflectivity, types of hydrometeors) are available at 1 km resolution and at a 5-minute frequency. Their assimilation in kilometric resolution models ensures a better representation of the variability of Mediterranean rainfall events [CAU 10, WAT 14].

Despite this, weather forecasts remain affected by uncertainties. In order to quantify these forecasting errors, ensemble forecasting systems are implemented. The idea is to describe the initial state of the atmosphere by several analyses, selected in such a way that they are consistent with the observations and accompanying uncertainties. Different parameter settings are also used for describing the physical processes of the model. Finally, for limited area models, several conditions are applied to the domain's boundaries. Several predictions are made with the model, extracting data from different analyses, parameter settings and lateral boundary conditions. We thus obtain a set of possible scenarios for the evolution of meteorological parameters in the future, which is called an ensemble forecast. An ensemble prediction system forecast based on the AROME [BOU 12] model, called AROME-EPS has been operational at Météo-France since 2017.

Numerical model outputs are then evaluated with the expertise of forecasters. The latter interpret the fields of forecasting models so as to communicate the future weather conditions to users, decision makers and to the general audience. If a dangerous weather situation in the following 24 hours is foreseen, alert maps are issued (Figure 3.10) twice a day so as to warn the population. Heavy rainfall and thunderstorms are among the type of weather conditions monitored in this setting. A four-color code qualifies risk and more accurate warning bulletins are issued in case of high probabilities of severe weather (orange or red levels of alertness). These bulletins provide information about the current phenomenon, its expected evolution, its possible consequences and the potential damage it may cause, and guidelines to follow.

---

11 The troposphere is the atmosphere's layer where meteorological phenomena take place. It spreads from the earth's surface up to approximately 15 km altitude. In the first troposphere kilometers, meteorological phenomena are heavily influenced by the interactions between the atmosphere and the earth's surface.

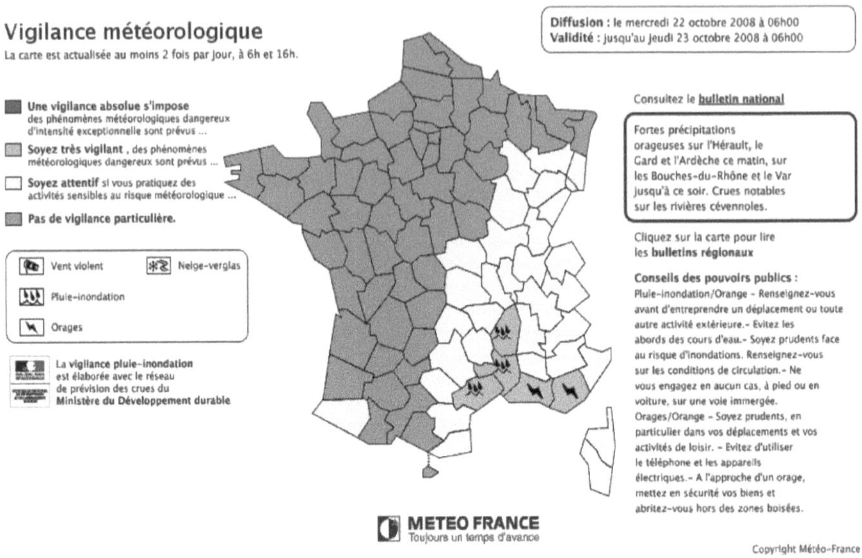

Figure 3.10. *Example of a meteorological Watch map for a MoBiClimEx case study (published on October 22, 2008). For a color version of this figure, see www.iste.co.uk/lutoff/hydrometeo1.zip*

3.3.2. *Flash flood forecasting methods*

The Flash Flood Guidance (FFG) method [GEO 86] was one of the first methods developed for the prediction of flash floods. The method relies on rainfall information and initial moisture conditions. Rainfall thresholds are defined in such a way that the flow rate calculated because of the rainfall amount corresponds to the overflow rate of the watercourse. Douinot *et al.* [DOU 16] evaluated this method in three French Mediterranean basins and tried to integrate the spatial variability of rainfall. Although they obtained encouraging results, these still need to be confirmed by more case studies. In France, warnings based on rain thresholds are carried out on the basis of the APIC[12] warning system (Intense Rain Warning for Municipalities), which was set up in 2011 by Météo-France [CAR 11].

Gourley *et al.* [GOU 14] assessed the contribution of meteorological radar data for flash flood forecasting in the United States. In comparison with the FFG method, the authors showed improvement in predictions when using a spatially distributed hydrological model, fueled by input radar data and then used in prediction mode.

12 *Avertissement Pluies Intenses à destination des Communes.*

Barthold *et al.* [BAR 15] evaluated how much earlier the forecasting of U.S. flash floods could be anticipated by using quantitative precipitation forecasts.

Some authors have explored how the uncertainties related to these rain forecasts (intensity, location of the most intense cells) can be taken into account in forecast chains leading to discharge ensemble forecasts. For example, Vincendon *et al.* [VIN 11] or Hardy *et al.* [HAR 16] used a perturbation method applied to the deterministic forecast of rainfall field that takes these errors into account. Amengual *et al.* [MEA 15] and Ravazzani *et al.* [RAV 16] showed the benefit of ensemble weather forecasts obtained by the Weather Research and Forecasting (WRF) model, fueled by an ensemble forecast provided by the European Center for Medium-Range Weather Forecasts[13]. The IMPRINTS project has also shown the added value of ensemble rainfall forecasts [ALF 11, WHO 12, LEL 13, VEL 13]. Tao *et al.* [TAO 16] evaluated (almost in real time) contributions to the quantitative precipitation forecast of a denser network of pluviographs and radars, as well as the interest of these rain estimates for flood forecasting. They were able to compare different rain products, different methods of data assimilation and they showed that the assimilation of discharge data led to very promising results. Tekeli *et al.* [TEK 16] showed the interest, in terms of alert, of a satellite estimate of rainfall, for countries where field measurements are rarely available, as in Saudi Arabia for example.

In terms of hydrological modeling tools, several forecasting chains are now available. Nguyen *et al.* [NGU 16] designed a semidistributed model, coupled with a hydraulic model, for simulating not only discharges, but also water levels and water velocities. This model is implemented with parameter values defined *a priori*, drawn from the literature and already acquired knowledge, but it is not calibrated. A sensitivity analysis, conducted with simulated rainfall fields, showed acceptable performance for an uncalibrated model. Artinyan *et al.* [ART 16] introduced a flash flood forecasting system, which is operational on the Arda cross-border basin between Bulgaria and Greece. This system is based on a calibrated distributed hydrologic model (ISBA-TOP, presented in section 3.3.4) where the flow routing module in the river has been improved. This system relies on real-time rain gauge data. The systems available in France are presented in the following section.

3.3.3. *Flash flood forecasting tools currently available in France*

In France, several flash flood forecasting modeling tools have been developed in partnership with Schapi: the MARINE, ATHYS, ALTHAIR and AIGA methods, among others. Here, we will offer an overview of their operation and state of development.

---

13 CEPMMT, *Centre Européen pour les Prévisions Météorologiques à Moyen Terme.*

The MARINE[14] (*Modélisation et Anticipation du RuIssellement pour des évèNements Extrêmes* [ROU 11]) model is an event-based distributed model, specifically dedicated to the simulation of Mediterranean floods. It is based on a square grid that corresponds to the digital terrain model (typically of a 25 meter resolution) and represents the following processes: runoff by infiltration capacity excess, subsurface lateral flows and flows in the hydrographic network, by approximation of the kinematic wave equation. For a given watershed, the model is generally calibrated using a set of flood events. The saturation state of the tanks is initialized using the outputs of the SIM model (SAFRAN-ISBA-MODCOU [HAB 08]), which runs continuously throughout France.

The event, distributed and conceptual SCS-LR model ([TRA 11], among others) developed by the IRD at Hydrosciences Montpellier and incorporated into the ATHYS[15] platform, combines a rainfall/runoff Soil Conservation Service (SCS) module, made up of a cascade of linear reservoirs, with the Lag and Route (LR) routing model. Catchments are discretized in square grids and on each of these grids, the runoff coefficient varies temporally, according to the cumulated rainfall and the initial water deficit on the grid. The LR routing module uses a propagation time and a diffusion time at each point of the grid.

Another "reservoir" model, ALHTAIR, has been developed by the SPC Grand Delta for forecasts in the Cévennes catchments [BRE 02]. The first ALHTAIR reservoir is called "pre-imbibation" and pours into an infiltration tank. When the volume of water in the latter reaches its maximum, the excess water constitutes the subsurface flow that will contribute to the outlet flow. This production function is applied in each mesh of the grid that describes the watershed. A transfer function is then applied to the flow produced by each mesh until the outlet is reached. For this, transfer time is calculated according to the geomorphological characteristics of the watershed.

Javelle *et al.* [JAV 14, JAV 16b] introduced the basic of the AIGA[16] method (see [LAV 05] for a first version of this method), which formed the basis for the *VigieFlash* system, operational since the beginning of 2017 on the French territory, and which made it possible to provide alerts at a municipal level. This method was particularly developed for the prediction of flash floods in watersheds of a few hundred km^2 and is based on a hydrological event-based model with one reservoir, with a kilometer grid that is fed by radar rainfall at a 15-min resolution. A daily hydrological model is used for initializing the reservoir level. The model provides simulations for river reaches corresponding to subbasins with a minimum size of 5 km^2.

---

14 MARINE, Modeling and Runoff Anticipation for Extreme Events.

15 *ATelierHYdrologique Spatialisé*, http://www.athys-soft.org/.

16 Adaptation from the Geographic Information for Flood Alerts: http://www.irstea.fr/la-recherche/unites-de-recherche/ohax/hydrologie/aiga.

In real time, the model is fueled by radar images at 15-min intervals and provides a flow value for each river reach. This value is compared with estimated quantiles for the whole of France that were estimated from a climatology of rainfall composed of the COMEPHORE radar/rain gauges reanalyses for the 1998–2006 period [GUE 11] and hourly ANTILOPE[17] radar images for the 2007–2014 period [JAV 16b]. If the value of the flow exceeds the quantiles, the river reach is colored according to the following scale:

– green: return[18] period <2 years;

– yellow: return period between 2 and 10 years;

– orange: return period between 10 and 20 years;

– red: return period between 20 and 50 years;

– purple: return period >50 years.

3.3.4. *Example of the result of a forecasting chain*

A flow forecasting chain made up of precipitation forecasts used as input to a hydrological model (as described in section 3.2.1) was designed and evaluated in the Cévennes catchment for the event of October 21 and 22, 2008, which was chosen as a case study for the MobiClimEx project [VIN 11].

Precipitation forecasts were provided by the AROME numerical weather prediction model in its ensemble version (AROME-EPS) and in its deterministic version. In the latter, a rainfall field perturbation method is applied for accounting for the uncertainty of these forecasts. The method is based on an analysis of previous errors committed within the AROME model. Different scenarios of precipitation can thus be obtained (Figure 3.11).

The hydrological model used is the ISBA-TOP system [BOU 09, WIN 10, WIN 16]. This model simulates Mediterranean flash floods taking advantage of the ISBA land surface scheme [NOI 89] so as to manage the soil–atmosphere interface. Hydrological processes are simulated on grids representing vertical transfers between the soil, vegetation and atmosphere (soil columns). The lateral distribution of water in the soil is carried out using topographic information as well as data concerning the distribution of rainfall within the watersheds.

---

17 http://wikhydro.developpement durable.gouv.fr/index.php/B.04_Estimation_d'une_pluie_de_bassin_par_observation_RADAR.

18 A T flood return period has one chance in T of taking place during a certain year.

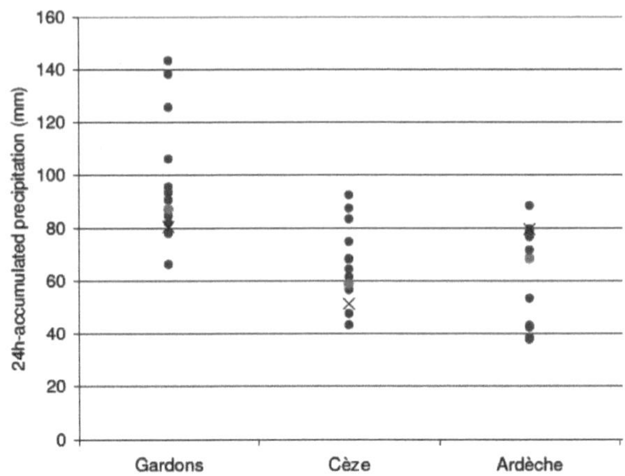

Figure 3.11. *Twenty-four hour rainfall accumulation (in mm) from October 21, 2008 at 12 UTC to October 22, 2008 at 12 UTC, average calculation for catchments, the Gardon at Ners, the Cede at Bagnold/Cede and the Ardèche at Vallon Pont d'Arc, drawn from radar observations (black crosses) from the AROME deterministic forecast model (pink dots) and the different members from AROME-EPS (blue dots). For a color version of this figure, see www.iste.co.uk/lutoff/hydrometeo1.zip*

The rainfall forecasting scenarios used as input for the ISBA-TOP make it possible to obtain several discharge scenarios, labelled discharge "ensemble" (Figure 3.12).

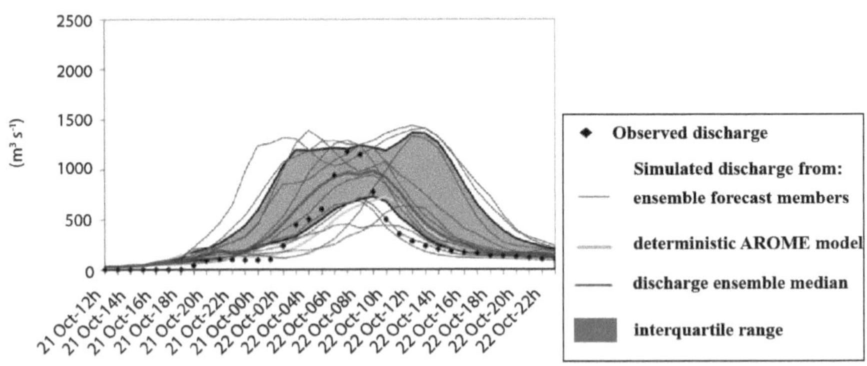

Figure 3.12. *Hourly discharges (in $m^3 \cdot s^{-1}$) of the Gardon at Ners from October 21, 2008 at 12 UTC to October 22, 2008 at 23 UTC observed (black rectangles) and simulated from the deterministic AROME forecast model (green curve) and AROME-EPS scenarios (blue curves). The discharge ensemble median is outlined in red, and the interquartile space is represented by the shadowed area in gray. For a color version of this figure, see www.iste.co.uk/lutoff/hydrometeo1.zip*

These discharge forecasts inform the hydrological forecaster about the probability of exceeding a certain discharge threshold, which may involve not only warning thresholds, but also the most probable flood peak time and the recession dynamics. This forecast could therefore be made with a 24-h anticipation, which can prove invaluable for putting security services on alert.

Using the simulation results of the ISBA-TOP model, quantile exceedance maps can be plotted (Figure 3.13), but unlike AIGA, these quantile values are based on the observed discharges. If obtained at the end of the hydrometeorological forecast chain, this type of information can sensibly help with decision making. In Figure 3.13, discharge forecasts are compared to road closure incidents, what makes it possible to anticipate flash flood impacts, as well.

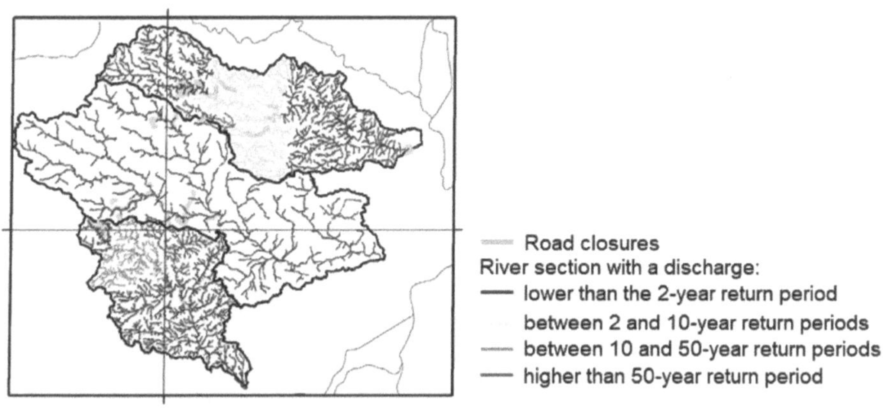

Figure 3.13. *Example of quantile exceedance for the case of October 2008. The simulated discharge for each section of the river is classified according to return periods of 3, 5, 10, and 50 years and a color is associated with each section depending on the discharge type (beware that the color code is not the same as for the AIGA described in section 3.3.3). Effective road closures are shown in green. Observe that this map was obtained by using radar data as input to the ISBA-TOP. For a color version of this figure, see www.iste.co.uk/lutoff/hydrometeo1.zip*

3.3.5. *Toward a flash flood impact forecast?*

The most recent research intends to go beyond exclusive discharge forecast and simulate flash flood impact. This implies adding an impact prediction model to the discharge forecasting chain previously described.

Until now, research focused on the prediction of road closures, with developments in the Gard department where data concerning road closures were available [VER 10, NAU 13, ANQ 15, WIN 16]. A vulnerability model of road sections is necessary, but this vulnerability is difficult to estimate [JAV 16a]. Le Bihan *et al.* [LEB 16a] extended this approach to an impact model, representing the number of potentially flooded buildings, by coupling the hydrological model (CINECAR) with a simplified hydraulic model, so as to obtain an estimate of water levels. When these projections are transferred to a digital terrain model, it can help to estimate the flooded areas. Debionne *et al.* [DEB 16] proposed a traffic model that simulates the number of vehicles on a road section during the day. When this information is crossed with a discharge simulation model, and road closures, it is possible to estimate the driver population exposed to road flooding [SHA 17].

Via You Tube, and concerning the south of France, Saint Martin *et al.* [SAI 16] collected web or social network information about the impacts of several recent damaging flash floods. These data were used for assessing the relevance of the alerts provided by the AIGA method. Work in progress aims at refining the AIGA method, by taking into account major stakes and possible damage, in order to target alerts on problematic areas more efficiently.

More generally, the prediction of flood impacts must account for the interactions between the dynamics of society's vulnerabilities [TER 15] and natural phenomena, at different scales of time and space, which call for innovative observation strategies (crowd-sourcing, social networks, etc.).

3.4. Conclusion

Predicting Mediterranean flash floods requires paying attention to spatial scales of the order of a few km^2, and to time scales ranging from a few minutes to a few hours, depending on the generating type of rainfall event. These floods affect densely populated ares, either permanently or during the tourist season. These areas are often regions with a high socioeconomic activity. They can therefore cause a lot of material damage and even be lethal. Therefore, the need to predict them more accurately and with enough lead time to favor a suitable social response is crucial, but their violence and speed make them difficult to predict.

Recent progress has made it possible to adapt the different stages of the forecasting chain to the appropriate flash flood scales. First, current methods of rain observation (especially those based on meteorological radars) have favored an accurate spatiotemporal description of Mediterranean rainfalls. Progress has also been made on numerical weather prediction models. Their resolution is now sufficiently high so as to realistically predict the precipitation fields generated by Mediterranean rainfall systems. Furthermore, methods have been developed to associate these forecasts with information about their uncertainty. Finally, in the last few years, hydrological models have been substantially improved. Now, they are spatially distributed so as to better describe catchments and to consider small-scale processes for simulating discharges in ungauged watersheds.

Other steps can be added to these forecasting chains in order to provide information that is not only hydrometeorological, but also concerns the impact of floods, such as road network closures and areas of potential damage.

The challenge for the next few years will be to create integrated forecasting chains comprising meteorology, hydrology, hydraulics and social dynamics, so as to anticipate the possible consequences of floods. The following chapters will aim to provide some key elements in order to move forward in this direction.

3.5. Bibliography

[AHR 17] AHROUCH S., Évaluation de modèles pour la prévision des crues sur le bassin de l'Yzeron (sud-ouest de Lyon), End of study project, ENS3, Grenoble, 2017.

[ALF 11] ALFIERI L., SMITH P.J., THIELEN-DEL POZO J. *et al.*, "A staggered approach to flash flood forecasting case study in the Cévennes region", *Advances in Geosciences*, vol. 29, pp. 13–20, 2011.

[AME 15] AMENGUAL A., HOMAR V., JAUME O., "Potential of a probabilistic hydrometeorological forecasting approach for the 28 September 2012 extreme flash flood in Murcia, Spain", *Atmospheric Research*, vol. 166, pp. 10–23, 2015.

[ANQ 15] ANQUETIN S., VANNIER O., OLLAGNIER M. *et al.*, "Flash-flood hydrological simulations at regional scale.Scale signature on road flooding vulnerability", *EGU General Assembly*, April 12–17, 2015, Vienna, Austria, 2015.

[ART 16] ARTINYAN E., VINCENDON B., KROUMOVA K. *et al.*, "Flood forecasting and alert system for Arda River basin", *Journal of Hydrology*, vol. 541, pp. 457–470, 2016.

[BAR 15] BARTHOLD F.E., WORKOFF T.E., COSGROVE B.A. *et al.*, "Improving flash flood forecasts the HMT-WPC flash flood and intense rainfall experiment", *Bulletin of the American Meteorological Society*, vol. 96, no. 11, pp. 1859–1866, 2015.

[BOR 10] BORGA M., ANAGNOSTOU E.N., BLÖSCHL G. *et al.*, "Flash floods: observations and analysis ofhydrometeorological controls", *Journal of Hydrology*, vol. 394, nos 1–2, pp. 1–3, 2010.

[BOR 12] BORGA M., ANAGNOSTOU E.N., BLÖSCHL G. *et al.*, "Flash flood forecasting, warning and risk management: the HYDRATE project", *Environmental Science & Policy*, vol. 14, no. 7, pp. 834–844, 2012.

[BOU 09] BOUILLOUD L., CHANCIBAULT K., VINCENDON B. *et al.*, "An advanced coupling between the ISBA land surface model and the TOPMODEL hydrological model to simulate Mediterranean flash-floods", *Journal of Hydrometeorology*, vol. 11, no. 2, pp. 315–333, 2009.

[BOU 12] BOUTTIER F., VIÉ B., NUISSIER O. *et al.*, "Impact of stochatic physics in a convection-permitting ensemble", *Monthly Weather Review*, vol. 140, pp. 3706–372, 2012.

[BOU 16] BOUDEVILLAIN B., DELRIEU G., WIJBRANS A. *et al.*, "A high-resolution rainfall re-analysis based on radar–raingauge merging in the Cévennes-Vivarais region, France", *Journal of Hydrology*, vol. 541, pp. 14–23, 2016.

[BRA 14] BRAUD I., AYRAL P.A., BOUVIER C. *et al.*, "Multi-scale hydrometeorological observation and modelling for flash flood understanding", *Hydrology and Earth System Sciences*, vol. 18, no. 9, pp. 3733–3761, 2014.

[BRA 16] BRAUD I., ANDRIEU J., AYRAL P.-A. *et al.*, FloodScale: multi-scale hydrometeorological observation and modelling for flash floods understanding and simulation, End of project report, ANR- 2011 BS56 027 01, Project, 2016.

[BRE 02] BRESSAN, F., " Le projet ALHTAIR du service d'annonce des crues du Gard ", *La Houille Blanche*, vol. 2, pp. 64–68, 2002.

[BRO 16] BROUSSEAU P., SEITY Y., RICARD D. *et al.*, "Improvement of the forecast of convective activity from the AROME-France system", *Quarterly Journal of the Royal Meteorological Society*, vol. 142, no. 699, pp. 2231–2241, 2016.

[CAR 11] CARRIERE J.-M., BROVELLI P., TZANOS R., "Apic: service d'avertissement de pluie intense ", *Météo Le Magazine*, vol. 14, pp. 18–19, 2011.

[CAU 10] CAUMONT O., DUCROCQ V., WATTRELOT É. *et al.*, "1D+3DVar assimilation of radar reflectivity data: a proof of concept", *Tellus A*, vol. 62, no. 2, pp. 173–187, 2010.

[CRE 03] CREUTIN J.-D., BORGA M., "Radar hydrology modifies the monitoring of flash-flood hazard", *Hydrological Processes*, vol. 17, no. 7, pp. 1453–1456, 2003.

[DEB 16] DEBIONNE S., RUIN I., SHABOU S. *et al.*, "Assessment of commuters' daily exposure to flash flooding over the roads of the Gard region, France", *Journal of Hydrology*, vol. 541, pp. 636–648, 2016.

[DEL 05] DELRIEU G., DUCROCQ V., GAUME E. *et al.*, "The catastrophic flash-flood event of 8-9 September 2002 in the Gard region, France: a first case study for the Cévennes-Vivarais Mediterranean hydrometerorological observatory", *Journal of Hydrometeorology*, vol. 6, no. 2, pp. 34–52, 2005.

[DEL 14] DELRIEU G., WIJBRANS A., BOUDEVILLAIN B. *et al.*, "Geostatistical radar–raingauge merging: a novel method for the quantification of rain estimation accuracy", *Advances in Water Resources*, vol. 71, pp. 110–124, 2014.

[DEL 16] DELRIEU G., RUIN I., GAUME E. *et al.*, "Interdisciplinary post-event surveys to disentangle hazard from vulnerability in the impacts of Mediterranean flash-floods events", *The Mediterranean Region under Climate Change. A Scientific Update*, Editions IRD, Allenvi, 2016.

[DOU 16] DOUINOT A., ROUX H., GARAMBOIS P.-A. *et al.*, "Accounting for rainfall systematic spatial variability in flash flood forecasting", *Journal of Hydrology*, vol. 541, pp. 359–370, 2016.

[DRO 14] DROBINSKYI P., DUCROCQ V., ALPERT P. *et al.*, "HyMeX, a 10-year multidisciplinary program on the Mediterranean water cycle", *Bulletin of the American Meteorological Society*, vol. 95, no. 7, pp. 1063–1082, 2014.

[DUC 02] DUCROCQ V., RICARD D., LAFORE J.-P. *et al.*, "Storm-scale numerical rainfall prediction for five precipitating events over France: on the importance of the initial humidity field", *Weather and Forecasting*, vol. 17, pp. 1236–1256, 2002.

[GAU 08] GAUME E., BORGA M., "Post-flood field investigations in upland catchments after major flash floods: proposal of a methodology and illustrations", *Journal of Flood Risk Management*, vol. 1, no. 4, pp. 175–189, 2008.

[GAU 09] GAUME E., BAIN V., BERNARDARA P. *et al.*, "A compilation of data on European flash floods", *Journal of Hydrology*, vol. 367, nos 1–2, pp. 70–78, 2009.

[GEO 86] GEORGAKAKOS K.P., "On the design of national, real time warning systems with capability for se-specific flash flood forecasts", *Bulletin of the American Meteorological Society*, vol. 67, pp. 1233–1239, 1986.

[GOU 14] GOURLEY J. J., FLAMIG Z.L., HONG Y. *et al.*, "Evaluation of past, present, and future tools for radar-based flash flood prediction", *Hydrological Sciences Journal,* vol. 59, pp. 1377–1389, 2014.

[GUE 11] GUEGUEN C., L'HENAFF G., LAURANTIN O. *et al.*, *Description des traitements en vue d'une réanalyse des lames d'eau*, Météo-France, 2011.

[HAB 08] HABETS F., BOONE A., CHAMPEAUX J.L. *et al.*, "The hydrometerological model applied over France", *Journal of Geophysical Research*, vol. 113, p. D06113, 2008.

[HAR 16] HARD, J., GOURLEY J.J., KIRSTETTER P.-E. *et al.*, "A method for probabilistic flash flood forecasting", *Journal of Hydrology*, vol. 541, pp. 480–494, 2016.

[HAU 08] HAUET A., KRUGER A., KRAJEWSKI W.F. *et al.*, "Experimental system for real-time discharge estimation using an image-based method", *Journal of Hydrologic Engineering*, vol. 13, no. 2, pp. 105–110, 2008.

[JAV 14] JAVELLE P., DEMARGNE J., DEFRANCE D. *et al.*, "Evaluating flash-flood warnings at ungauged locations using post-event surveys: a case study with the AIGA warning system", *Hydrological Sciences Journal*, vol. 59, no. 7, pp. 1390–1402, 2014.

[JAV 16a] JAVELLE P., BRAUD I., SAINT MARTIN C. *et al.*, "Improving flash flood forecasting and warning capabilities", *The Mediterranean Region under Climate Change. A Scientific Update*, Editions IRD, Allenvi, 2016.

[JAV 16b] JAVELLE P., ORGANDE D., DEMARGNE J. *et al.*, "Setting up a French national flash flood warning system for ungauged catchments based on the AIGA method", *FLOODrisk 2016 – 3rd European Conference on Flood Risk Management*, Lyon, France, 2016.

[LAV 05] LAVABRE J., GREGORIS Y., "AIGA: un dispositif d'alerte des crues sur l'ensemble du réseau hydrographique", *Ingénieries EAT*, vol. 44, pp. 3–12, 2005

[LEB 16a] LE BIHAN G., PAYRASTRE O., GAUME E. *et al.*, "Regional hydrological models for distributed flash-floods nowcasting: towards an estimation of potential impacts and damages", *Proceedings of the 3rd European Conference on Flood Risk Management*, Lyon, France, October 17–21, 2016.

[LEB 16b] LE BOURSICAUD R., PÉNARD L., HAUET A. *et al.*, "Gauging extreme floods on YouTube: application of LSPIV to home movies for the post-event determination of stream discharges", *Hydrological Processes*, vol. 30, no. 1, pp. 90–105, 2016.

[LEC 10] LE COZ J., HAUET A., PIERREFEU G. *et al.*, "Performance of image-based velocimetry (LSPIV) applied to flash-flood discharge measurements in Mediterranean rivers", *Journal of Hydrology*, vol. 394, nos 1–2, pp. 42–52, 2010.

[LEC 16] LE COZ J., PATALANO A., COLLINS D. *et al.*, "Crowdsourced data for flood hydrology: feedback from recent citizen science projects in Argentina, France and New Zealand", *Journal of Hydrology*, vol. 541, pp. 766–577, 2016.

[LIE 13] LIECHTI K., PANZIERA L., GERMANN U. *et al.*, "The potential of radar-based ensemble forecasts for flash-flood early warning in the southern Swiss Alps", *Hydrology and Earth System Sciences*, vol. 17, pp. 3853–3869, 2013.

[MAR 10] MARCHI L., BORGA M., PRECISO E. *et al.*, "Characterisation of selected extreme flash floods in Europe and implications for flood risk management", *Journal of Hydrology*, vol. 394, nos 1–2, pp. 118–133, 2010.

[NAU 13] NAULIN J.P., PAYRASTRE O., GAUME E., "Spatially distributed flood forecasting in flash flood prone areas: application to road network supervision in Southern France", *Journal of Hydrology*, vol. 486, pp. 88–99, 2013.

[NGU 16] NGUYEN P., THORSTENSEN A., SOROOSHIAN S. *et al.*, "A high resolution coupled hydrologic– hydraulic model (HiResFlood-UCI) for flash flood modeling", *Journal of Hydrology*, vol. 541, pp. 401–420, 2016.

[NOI 89] NOILHAN J., PLANTON S., "A simple parametrization of land surface processes for meteorological models", *Monthly Weather Review*, vol. 117, pp. 536–549, 1989.

[QUI 12] QUINTERO F., SEMPERE-TORRES D., BERENGUER M. *et al.*, "A scenario-incorporating analysis of the propagation of uncertainty to flash flood simulations", *Journal of Hydrology*, vol. 460, pp. 90–102, 2012.

[RAV 16] RAVAZZANI G., AMENGUAL A., CEPPI A. *et al.*, "Potentialities of ensemble strategies for flood forecasting over the Milano urban area", *Journal of Hydrology*, vol. 539, pp. 237–253, 2016.

[ROU 11] ROUX H., LABAT D., GARAMBOIS P.A. *et al.*, "A physically-based parsimonious hydrological model for flash floods in Mediterranean catchments", *Natural Hazards and Earth System Sciences*, vol. 11, no. 9, pp. 2567–2582, 2011.

[RUI 08] RUIN I., CREUTIN J.-D., ANQUETIN S. *et al.*, "Human exposure to flash-floods-relation between flood parameters and human vulnerability during a storm of September 2002 in Southern France", *Journal of Hydrology*, vol. 361, pp. 199–213, 2008.

[RUI 14] RUIN I., LUTOFF C., BOUDEVILLAIN B. *et al.*, "Social and hydrological responses to extreme precipitations: an interdisciplinary strategy for postflood investigation", *Weather Climate and Society*, vol. 6, no. 1, pp. 135–153, 2014.

[SAI 16] SAINT-MARTIN C., FOUCHIER C., JAVELLE P. *et al.*, "Assessing the exposure to flooding to implement an flood impact model for French Mediterranean basins", *Proceedings of the 3rd European Conference on Flood Risk Management*, Lyon, France, October 17–21, 2016.

[SEI 11] SEITY Y., BROUSSEAU P., MALARDEL S. *et al.*, "The AROME-France convective scale operational model", *Monthly Weather Review*, vol. 139, pp. 976–991, 2011.

[SHA 17] SHABOU S., RUIN I., LUTOFF C. *et al.*, "MobRISK: a model for assessing exposure of road users to flash flood events", *Natural Hazards and Earth System Sciences*, vol. 17, pp. 1631–1651, 2017.

[TAB 07a] TABARY P., "The new French radar rainfall product. Part I: methodology", *Weather and Forecasting*, vol. 22, no. 3, pp. 393–408, 2007.

[TAB 07b] TABARY P., DESPLATS J., DO KHAC K. *et al.*, "The new French radar rainfall product. Part II: Validation", *Weather and Forecasting*, vol. 22, no. 3, pp. 409–427, 2007.

[TAO 16] TAO J., WU D., GOURLEY J. *et al.*, "Operational hydrological forecasting during the IPHEx-IOP campaign – meet the challenge", *Journal of Hydrology*, vol. 541, pp. 434–456, 2016.

[TEK 16] TEKELI A.E., FOULI H., "Evaluation of TRMM satellite-based precipitation indexes for flood forecasting over Riyadh City, Saudi Arabia", *Journal of Hydrology*, vol. 541, pp. 471–479, 2016.

[TER 15] TERTI G., RUIN I., ANQUETIN S. *et al.*, "Conceptualizing dynamic social vulnerability. Space-time factors considering flash flood events", *Natural Hazards*, vol. 79, pp. 1481–1497, 2015.

[THI 08] THIELEN J., BARTHOLMES J., RAMOS M.-H. *et al.*, "The European flood alert system – part 1: concept and development", *Hydrology and Earth System Sciences Discussions*, vol. 5, no. 1, pp. 257–287, 2008.

[TRA 11] TRAMBLAY Y., BOUVIER C., AYRAL P. *et al.*, "Impact of rainfall spatial distribution on rainfall runoff modeling efficiency and initial soil moisture conditions estimation", *Natural Hazards and Earth System Sciences*, vol. 11, pp. 157–170, 2011.

[VEL 13] VELASCO M., VERSINI P.A., CABELLO A. *et al.*, "Assessment of flash floods taking into account climate change scenarios in the Llobregat River basin", *Natural Hazards and Earth System Sciences*, vol. 13, pp. 3145–3156, 2013.

[VER 10] VERSINI P.A., GAUME E., ANDRIEU H., "Assessment of the susceptibility of roads to flooding based on geographical information – test in a flash flood prone area (the Gard region, France)", *Natural Hazards and Earth Systems Sciences*, vol. 10, pp. 793–803, 2010.

[VIN 07] VINET F., "Approches nationales de la prévention des risques et besoins locaux: le cas de la prévision et de l'alerte aux crues dans le midi méditerranéen", *Géocarrefour*, vol. 82, nos 1–2, pp. 35–42, 2007.

[VIN 10] VINCENDON B., DUCROCQ V., SAULNIER G.M. *et al.*, "Advantages of coupling the ISBA land surface model with a TOPMODEL hydrological model dedicated to Mediterranean flash floods", *Journal of Hydrology*, vol. 394, nos 1–2, pp. 256–266, 2010.

[VIN 11] VINCENDON B., DUCROCQ V., NUISSIER O. *et al.*, "Perturbation of convection-permitting NWP forecasts for flash-flood ensemble forecasting", *Natural Hazards and Earth System Sciences*, vol. 11, no. 5, pp. 1529–1544, 2011.

[VIN 16] VINCENDON B., ÉDOUARD S., DEWAELE H. *et al.*, "Modeling flash floods in southern France for road management purposes", *Journal of Hydrology*, vol. 541, pp. 190–205, 2016.

[WAT 14] WATTTRELOT E., CAUMONT O., MAHFOUF J. F., "Operational implementation of the 1D+ 3D-Var assimilation method of radar reflectivity data in the AROME model", *Monthly Weather Review*, vol. 142, no. 5, pp. 1852–1873, 2014.

[WEL 16] WELBER M., LE COZ J., LARONNE J. *et al.*, "Field assessment of non-contact stream gauging using portable surface velocity radars (SVR)", *Water Resources Research*, vol. 52, no. 2, pp. 1108–1126, 2016.

[YOU 08] YOUNIS J., ANQUETIN S., THIELEN J., "The benefit of high-resolution operational weather forecasts for flash-flood warning", *Hydrology and Earth System Sciences*, vol. 12, pp. 1039–1051, 2008.

4

High-impact Weather Events: Is a Socio-hydrometeorological Characterization Possible?

4.1. Introduction

Both at a national and at a global level [BEN 16], the risk of flooding represents the most important risk factor of natural origin. In a context where one of the effects of global warming could lead to an intensification of the hydrological cycle (for instance, through an increase in the frequency and intensity of rainfall extremes and drought events [IPC 14]), it then becomes necessary to clarify the concept of "high-impact weather event". In specialized literature or in legislative texts, these events are characterized as "remarkable", "major", "extreme", or "foundational", and they are the targets for which territorial actors must develop management and anticipation tools, if they are to fulfill their missions.

This problem is not new. Chapter 5 will further explore how the evolution of national and local policy toward flooding has evolved since the 18th Century, following the rhythm of "events", all of which clearly produced high impacts. For instance, after the floods of the Seine and the Loire in 1910, which particularly affected Paris, many reflections were initiated on how to better manage the impact of such hazards. Nowadays, the economic cost (to name only one type of impact) of floods oscillates between 10 and 30 billion euros, a sum which would probably put the current CatNat compensation system [OEC 14] in a difficult position.

Chapter written by Sandrine ANQUETIN, Isabelle RUIN, Gilles MOLINIÉ, Jean-Stéphane CARNEL, Benoit LAFON, Isabelle BRAUD, Céline LUTOFF, Bruno WILHELM and Jean-Dominique CREUTIN.

In fact, the law, which regulates compensation for victims of disasters[1], ensures the compensation of any natural or legal person other than the State who has subscribed to an insurance contract against fire damage or any other damage to property located in France, thus guaranteeing the rights of the insured against the effects of natural disasters. In France, a natural disaster is characterized by an interministerial decree, which delimits affected areas as well as the duration of the disaster, and identifies the nature of the events and that of incurred damage. On the basis of certain criteria, it is a judge who must designate the event as a natural disaster. We can then consider that events identified as "natural disasters" under this legislation, and justifying State support compensating for damage, may be considered as high-impact weather events, at least for the most part. So, the question here is to establish the threshold after which an event can be understood as a high-impact weather event. Is this designation homogeneous? Does the designation include only those events which had an impact? What steps should be taken in order to formalize and to extend this definition beyond France's borders?

The Website GéoRisques[2] follows on Haroun Tazieff's comments: "The definition I give of major risk is a threat to man and his direct environment, to his buildings; a threat whose gravity is such that society is absolutely overwhelmed by the immensity of the disaster". A high-impact weather event can therefore be an event whose impact exceeds society's management and anticipation capacities.

By focusing this chapter on flash flood risk dynamics, the high-impact weather events we are interested in result from an unfavorable conjunction between (1) hydrometeorological hazards (for instance, rainfall intensity, flood dynamics and the extent of the flood) and (2) population vulnerability (for example, human and material exposure, perception and acceptability of risk, the ability of the exposed population to cope with the event). Offering a definition for such events requires mobilizing different kinds of knowledge so as to jointly characterize the severity of the hydrometeorological hazard and its associated socioeconomic impact, together with the corresponding spatiotemporal dynamic. This cognitive stage is essential for providing specific tools for forecasting, managing risk, predicting or producing scenarios and for spatial management. Nowadays, such systemic approaches are rare, although much progress has been made in recent years, particularly regarding

1 Law No. 82-600, from July 13, 1982, regulates compensation to victims from natural disasters. Its consolidated version of August 1, 2008 was resumed by article L-125-1, and modified by the law of December 25, 2007, from the French insurance code.

2 *Ministère de la Transition Ecologique et Solidaire* – http://www.georisques.gouv.fr/.

the understanding and modeling of (1) hydrometeorological hazards (for instance, [KHO 16] in the context of the international HyMeX program [DRO 14]) and (2) mobility against flash floods [SHA 17, TER 17]. In addition, Boudou [BOU 15] and Boudou *et al.* [BOU 17] suggested a reading grid for assigning a remarkability score to an event of climatic origin, combining information on the intensity of the hazard with its socioeconomic impact. For example, this grid has been applied to the events listed in the BDHI database, presented in the next section.

In general, the studies that aim at identifying the impact of hydrometeorological hazards define the "remarkable" nature of an event from its very beginning. For example, Météo-France or SCHAPI (*Service Central d'Hydrométéorologie et d'Appui à la Prévision d'inondation*[3]) rely on statistical definitions, respectively, based on the return periods of precipitation totals, as well as on observed or predicted flows. A return period exceeding 100 years represents an extreme event. On the basis of this information, forecasting services can therefore issue a graduated alert (yellow, orange, red) to inform and, if necessary, to mobilize the operational services and the population so as to deal with the potential effects of these hazards. The difficulty of this approach concerns (1) the "*a priori*" definition of the event, (2) the spatial resolution scale at which the vigilance bulletin is issued (departmental resolution for rainfall, river section resolution for flows) and (3) comments regarding possible consequences and suggested behavior to be adopted. Nowadays, information associated with vigilance bulletins in France remains rather uncertain and is not always in harmony with international standards.

Studies that seek to better identify the social impact of hazards tend to focus on exposure factors and on human and social vulnerability (e.g. [SHA 17, TER 17]). Exposure involves a spatial component that makes it possible to foresee the expected extension of the hazard. Exposure measurement involves an enumeration of buildings, resident populations and issues at stake in the flood zone. On the other hand, vulnerability focuses on the features of the sources of fragility and the ability to cope with them, depending on whether these are considered individually or collectively, in the short term (crisis) or in the long term (post- and intercrisis). *A priori*, the combined study of exposure and vulnerability makes it possible to consider the potential impact that a society could suffer if it was affected by a dangerous phenomenon of a certain spatial extent. While this generic approach is very useful for preventive purposes, it overlooks the dynamic components that simultaneously feature the hazard, the exposure and the vulnerability of individuals, in particular. Other approaches starting from the analysis of the "human impact" of flash floods have revealed the interest of simultaneously taking into account the

3 Central Hydrometeorology and Flood Forecasting Support Service.

spatiotemporal dynamics of the phenomenon and human vulnerability and exposure, which condition the spatiotemporal distribution of social impacts ([RUI 08, SHA 17, TER 16, TER 17]). This type of approach seems more appropriate for crisis forecasting and crisis management.

Here, we are measuring the difficulty of accurately defining high-impact weather events. In fact, this definition must necessarily be based on an understanding of the interrelationships between the hazard and its associated impacts, the latter being strongly linked to the affected populations and their ability to anticipate and react to the event. The need for an interdisciplinary perspective makes it necessary to clearly define the concepts to enhance comprehension. As can be appreciated in Figure 4.1, this interdisciplinary field is still widely open. For illustrative purposes, the figure presents word clouds from abstracts of a choice of recent papers (2010–2017), extracted from the international Web of Science database, using keywords related to the title "high-impact weather event". Forty publications were extracted and divided into two groups related to the disciplinary audience of the paper where the publications appeared (geosciences, human and social sciences). Thus, these two clouds of words suggested a "disciplinary" vision of the same object, in this case, high-impact events. The cross-analysis between these two types of terminology highlights the few words that the two communities have in common, beyond superficial similarities. While hydrometeorologists use "high-impact" terminology in order to indicate that they are working on a certain type of event (strong-extreme), they are careful to mention the socioeconomic potential echoes of their study, but they do not study impact *in fine*. Conversely, humanities and social sciences professionals employ the term "weather event" only to indicate the type of hazard at the origin of the impact they are studying.

Based on this analysis, we identified three issues that we will explore in this chapter. The first one concerns the scarcity of observations and necessary data for building a systemic approach for featuring high-impact weather events; this point will be discussed in section 2. Section 3 will propose an approach for characterizing a hazard depending on its spatial and temporal dimensions, all of which distinguish the "severity" of the event from a hydrometeorological point of view. Finally, vulnerability to these hazards is strongly linked to perception, acceptability, coping, and adaptation abilities of the population. To explore these aspects, section 4 will explore the media resonance of these events by comparing flash flood events of varying physical severity.

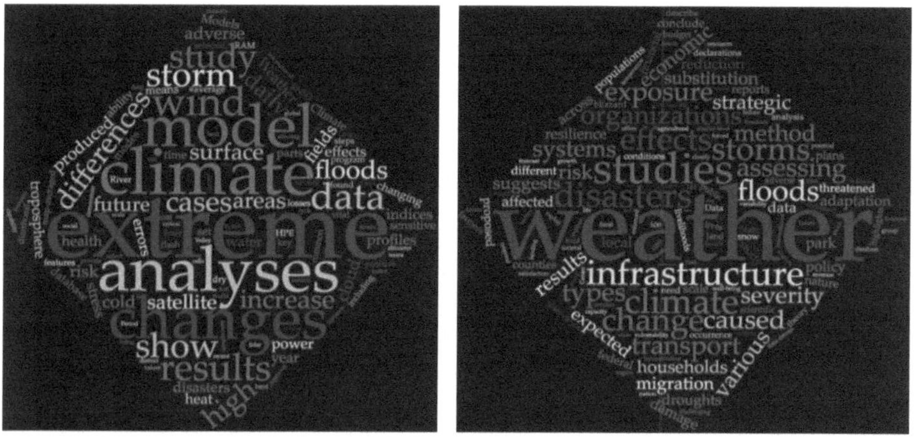

a) b)

Figure 4.1. *Word clouds extracted from the abstracts of a selection of research papers taken from the Web Of Sciences database, by using the title key words "high-impact weather event", for the period 2010–2017. Scientific papers were then classified in function of their audience and the issues aimed at: a) atmospheric and climatic science communities and b) human and social science communities, corresponding to around 20 papers for each community. For a color version of this figure, see www.iste.co.uk/lutoff/hydrometeo1.zip*

4.2. Available data for featuring the physical and socioeconomic impact of past events

The identification and analysis of sociohydrometeorological processes generating impact during flash flood events require the implementation of a systemic approach that mobilizes knowledge from the different "disciplines" involved (for example meteorology, hydrology, geography, sociology and economy) and associated data.

Therefore, the characterization of high-impact weather events requires data access, together with the expertise of the producers of disciplinary data, in order to specify the validity scope and the uncertainties associated with the collected data.

In this context, one of the challenges concerns the collection of multidisciplinary data, storage and a facilitated provision for scientific and/or operational communities, for these to be able to carry out their respective missions. Such databases are not numerous. This section will briefly introduce each of them and will apply their criteria to a case study, so as to highlight the limitations and uncertainties of the aforementioned databases.

4.2.1. *Existing databases*

Nowadays, we can identify five databases, which can meet multidisciplinary information collection needs (for example hazard characterization, information about associated economic and human impacts) in order to help with the definition of high-impact weather events (Table 4.1). These databases can be especially consulted for studies concerning the entire French national territory.

Name	Spatial resolution	Time frame	Type of event	Minimum membership criteria
EM-DAT	Country to department	1900–present day	Natural or industrial disaster	At least one criterion among: - at least 10 casualties - at least 100 affected or damaged people - emergency state declaration - call for international help
DFO	Country to municipality	1985–present day	Flooding	At least one criterion among: - Important damage to infrastructure or agriculture - Casualties
CATNAT	Country to department	2001–present day	Natural-origin destructive event	An event that engenders significant human or material consequences
FloodHymex	Country to administrative region	1981–2015	Flooding Flash flood Earthquake	Every measurable event
BDHI	Municipality	18th Century–present day	Flooding	Every noticeable event

Table 4.1. *Summary of the main features of databases associated with high-impact events. For further details, please refer to each of the corresponding Websites*

4.2.1.1. *International databases*

We can mention three international databases, which make information about high-impact weather events accessible to research.

The EM-DAT database (EMergencyDATa Base) was developed by the *Centre for Research on the Epidemiology of Disaster* at the Université Catholique de Louvain, Belgium. Built for the purpose of managing humanitarian interventions at international level, Guha-Sapir *et al.* [GUH 13] documented more than 20,400 disasters worldwide having taken place since 1900. Data archived in EM-DAT are derived from a variety of sources: the United Nations (UN), non-governmental organizations (NGOs), insurance companies and research institutes. EM-DAT identifies technological and natural disasters; among the latter, six types of origin can be distinguished: biological, climatological, extraterrestrial, geophysical, hydrological and meteorological[4].

The database kept by the University of Colorado's *Dartmouth Flood Observatory* (DFO) has archived data concerning world flood events since 1985. In addition to the actual data, DFO provides a map of floods obtained from satellite data. DFO collects information from the press, government agencies and *in situ* and satellite observation networks. Until recently, the database has documented more than 4,200 flood events and nearly 640,000 associated deaths. Information about casualties is often extracted from newspapers that specialize in the field, but relies on terminology that can poorly account as quantitative. For example, when the term "hundreds" was indicated in the original source, the DFO translated "300"; the term "scores", used in the database, was quantified with the value "30". Boissier [BOI 13] criticized this approach, since it introduces many sources of uncertainty. On the other hand, Jonkman [JON 05] pointed out several errors related to this classification, and suggested collecting more precise data, thus warning future database users about its limitations.

The CATNAT (*CATastrophes NATurelles* (Natural disasters)) database provides data characterizing events which have taken place since 2001; created by Ubyrisk Consultants, the database is regularly updated. Full access to information is reserved for subscribers (only the first five requested events can be accessible free of charge). As soon as human or material damage is involved during a natural disaster, a description of the event and its associated impact are recorded. A 6-level damage severity scale is employed for measuring both the human and material consequences

---

4 Meteorological risk concerns extreme atmosphere conditions that can last between some minutes and several days; weather risk is related to weather processes ranging from intraseasonal scales to multi-10-day estimates.

of the event. The classification criteria are based on those used by the French Ministry of Ecology and Sustainable Development as part of its annual ranking of damaging events of natural origin. Concerning human consequences, each level corresponds to the number of casualties (for example level 1: probably injured individuals, but their number is inferior to 10, no casualties; level 3: injured individuals, there might be between 3 and 50 casualties; level 6: injured and over 50,000 deaths). With regard to material consequences, the levels correspond to the degree of destruction and the spatial extent of the damage (for instance level 1: highly localized partial damage on some buildings (number of damaged buildings lower than 10); level 3: between 500 and 1000 damaged buildings, several departments affected; level 6: seriously damaged several thousand buildings, highly broad extension).

4.2.1.2. *Database at the level of the Mediterranean basin*

Within the context of the international HYMEX project (*Hydrological cycle in Mediterranean Experiment*), Llasat *et al.* [LLA 13] have developed the FloodHymex database, in operation since 2010, by listing every flood and flash flood event since 1981 in northwestern Mediterranean Europe (that is to say, Spain (Catalonia and Balearic Islands), Italy (Calabria) and France (Occitania and Provence-Alpes-Côte d'Azur)). Flash floods are identified because of the duration of the event, which must not exceed one day. Contrary to the international databases introduced in the previous section, FloodHymex provides hydrological data (for example, maximum and/or daily flow), meteorological data (that is to say, maximum precipitation over 24 h or during the entire event), as well as a census of different types of recorded socioeconomic impact (damage to roads, bridges, buildings, historical sites, etc.).

4.2.1.3. *Database at the French territory level*

The *Base de Données Historiques sur les Inondations*[5] (BDHI) was built in response to the *Directive Inondations* (Flood Directive) (Article R.566-1 of the French Environmental Code). Since 2015, a web application developed by the *Laboratoire d'Informatique de Grenoble* (Laboratory of Computer Science of Grenoble) [LAN 16] enabled public access to this database. Therein, the most remarkable flood events are listed and documented with the largest amount of information available, particularly in relation to their hydrometeorological characteristics and associated impacts. Some listed events date back to the 18th Century.

5 BDHI, Historical Flood Database (see http://bdhi.fr/appli/web/welcome).

4.2.1.4. *Summary of the different databases*

Table 4.1 summarizes the main characteristics of the databases described in previous paragraphs. This synthesis reveals the diversity of the "types" of events recorded and archived, as well as the spatial resolution at which each event is described. However, these elements do not make interdatabase cross-analyses any easier. That kind of analysis would make it possible to benefit from a set of remarkable situations worldwide and help us to sharply define the concept of high-impact weather events.

4.2.2. *High-impact weather events in south-east France: database crossed viewpoints*

Given the criteria used by each of the structures operating these databases, stored information is heterogeneous, and can therefore lead to a form of uncertainty about data. In order to analyze the different sources of uncertainty, databases are here explored in terms of the question of flash flood events that have occurred in the south east of France.

With regard to the study area targeted by the Mobiclimex project, the aim was to identify the signature of the different characteristics of databases (for instance, criteria used for event recording, description of the event and its temporality, stored information concerning associated impact) related to the knowledge that can be derived from high-impact weather events in this region.

This cross-view was implemented for the period 1985–2010, that is to say, the biggest common period between the different databases.

In what concerns the extraction criteria, the BD-FloodHymex and BDHI identify the "flash flood" or "rainfall flood event" fields, a designation defined by a peak "rising time" lower than 2 h, or between 2 and 6 h. The other databases were queried for "hydrology" and "meteorology" fields. Over the period of interest, the number of recorded events varies greatly from one database to another (Table 4.2), which constitutes a first source of uncertainty related to the very nature of the "event", as defined in each of the databases.

	EM-DAT	DFO	CATNAT	FloodHymex	BDHI
Events	24	23	213	27	8
Casualties	248	206	286	187	98

Table 4.2. *Number of events and victims associated with each flood or flash flood event recorded in different databases over the period 1985–2010 for the south-eastern region of France*

For example, CATNAT-DB lists all events involving human or material damage. These events can show highly different hydrometeorological intensities and/or impacts. As Table 4.2 indicates, the mere fact of being listed in the CATNAT-DB does not necessarily entail the "high-impact" character for the referenced event. In contrast, DFO-DB only identifies significant events, for which a "severity" indicator (ranging between 1 and 2) is suggested on the basis of structural damage (Table 4.3) and an equation measuring "danger" (M) (equation [4.1]). For the choice of flash flood events in the south east of France, in DFO-DB "Large Flood" and then "extreme event" were chosen. For EM-DAT-DB, "Flood" and "General Flood" were selected.

$$M = \log \text{(severity} \times \text{duration of the event on the affected area)} \quad [4.1]$$

The number of events extracted from the BDHI is the smallest; this arises from the selection criteria retained, in which only rainfall flood events (flood peak rising time between 2 and 6 h) and flash flood events (flood peak rising time lower than 2 h) were selected. Probably several events extracted from other databases are present in BDHI, but they must be referenced with other criteria. Considering the criteria used by the managers of EM-DAT-DB and DFO-DB (Table 4.1), which are probably the most restrictive ones, the number of events recorded must certainly be far lower than CATNAT-DB, but also slightly lower than FloodHymEX. However, we observe that the number of victims associated with these events is higher, especially when we compare data from EM-DAT and DFO with those from FloodHymex and BDHI. A cross-analysis of these databases shows, on the one hand, that recorded events are not all the same and, on the other hand, that for the same event appearing in one or another database, stored information is not exactly the same. This concerns not only the duration of the event (for example starting date, finishing date), but also the associated impact (for example number of victims, material damage).

Severity indicator	Class	Return duration (T)
1	Large flood events	$10 < T < 20$ years
1.5	Very large flood event	$20 < T < 100$ years Event extension must be higher than 5,000 km^2
2	Extreme event	$T > 100$ years

Table 4.3. *"Severity" scale as defined in DFO-DB*

Predicting high-impact weather events in order to better anticipate them, and then communicate about their effects accordingly, requires a better understanding of past events. Based on the information stored in databases, the first step is to associate a hazard with its correlative human impact so as to enable the identification of meteorological precursors for forecasting, expected effects and anticipation. This section shows that archives and associated documentation of past events are highly heterogeneous, depending on the institutions in charge of this task. Thus, in order to accurately define a high-impact weather event, database harmonization work nowadays appears as unavoidable.

4.3. Concept of hydrometeorological hazard severity

The previous section focused on the difficulty of defining impact events solely based on international and national databases. In this section, we propose a new "severity" indicator for measuring pluviometric hazards. Calculated on the basis of rainfall estimated by weather prediction models [OLS 17], this indicator can be used as a precursor for a targeted vigilance of affected populations, at the same time that it enables an intercomparison between different events, identifying a "severity" spatiotemporal scale.

It is interesting to observe that, over the last decade, the concept of rainfall hazard severity, initially designated as *high precipitation event* (HPE) and later as *high-impact weather event* (HIW) in scientific literature, relies on rainfall thresholds. The choice of threshold is made *a priori* on the basis of previous observations and largely depends on the climatic regions considered. For example, among the many publications on this subject, Papagiannaki *et al.* [PAP 13] associated each "impact" rainfall threshold ranging between values smaller than 60 mm daily (or smaller than 15 mm/h) with the lowest "impact", and values greater than 100 mm over 24 h (or more than 25 mm/h) with the most serious corresponding impact. The limits of this approach are twofold: one single piece of information concerning the rainfall threshold is not enough if the spatial extent is not specified; on the other hand, objective justifications for the given thresholds are rarely provided, which poses the question of how to extrapolate information into the future, when the effects of climate change are likely to intensify extreme phenomena, both in occurrence and in intensity. All these reasons led us to propose a severity indicator for the event, established on the basis of previous rainfall and taking into account the spatial dimension of the rainfall event. The extrapolation of this indicator into future climatic conditions remains the same as for all studies that assess the impact of global warming, which today rely on the strong hypothesis of the stationarity of planet Earth.

4.3.1. *The severity diagram*

The severity diagram was originally proposed by Ramos *et al.* [RAM 05] in order to quantify the dangerousness of Mediterranean storms as observed in the city of Marseille. This approach was extended to a regional scale, that of the Cévennes Vivarais, by Ceresetti *et al.* [CER 11]. This diagram (Figure 4.2) represents the occurrence frequency of extreme rainfall as a function of spatial and temporal scales. Thus, it makes it possible to identify the time (temporal rainfall cumulation) and space scales (rainfall spatial integration) which show maximum "severity" levels (maximum rainfall return time). The "severity" scale can be represented either by the return time of observed rainfall (Figure 4.2) or by maximum rainfall intensity (Figure 4.3).

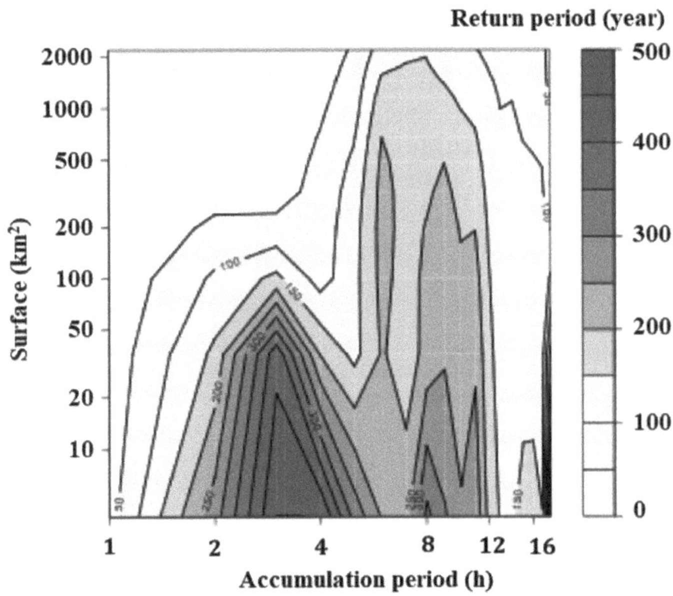

Figure 4.2. *Severity diagram. Iso-colors represent the rainfall return periods for the event considered. In this example, we can identify a strong severity for relatively weak time and space scales (e.g. 3–4 h; less than 100 km^2). For a color version of this figure, see www.iste.co.uk/lutoff/hydrometeo1.zip*

4.3.2. *Cross-analysis: severity diagram – socioeconomic impact*

Based on the events of the south-east region of France, taken from the different databases previously introduced, Granda [GRA 14] proposed a first study aimed at identifying the added value of severity diagrams to the construction of the concept of high-impact weather events. For this, the author developed severity diagrams for the 20 events identified in section 4.2 located in the region of Cévennes-Vivarais by crossing the information from the different databases for the period 1988–2008.

The 20 events were divided into four classes defined depending on the spatiotemporal footprint of the corresponding severity diagrams. Figure 4.3 shows an example of a representative diagram for each of these four classes.

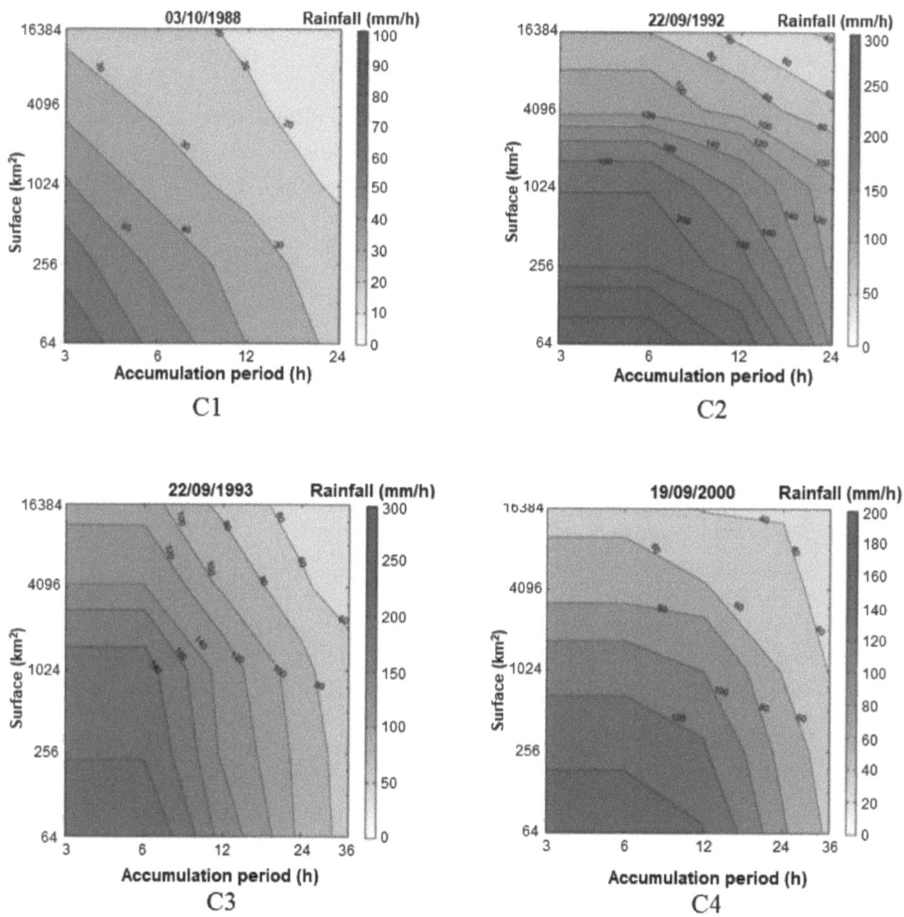

Figure 4.3. *Severity diagrams, expressing hourly maximal pluviographic intensity for four pluviographic events that affected the south east of France. These events are, respectively, associated with each of the types identified by Granda [GRA 14]. For a color version of this figure, see www.iste.co.uk/lutoff/hydrometeo1.zip*

Class 1 (Figure 4.3 C1; seven members) includes events for which severity related to observed maximum intensity is linked to the smallest time and space scales. Rainfall intensity is low. In this class, for instance, we can find the events that affected Nîmes, on October 3, 1988, and Hérault on December 1, 2003.

Class 2 (Figure 4.3 C2; four members) shows the highest rainfall intensities. The severity of this type of event affects important time and space scales. Therefore, events are more intense, last longer and have a larger spatial extent. In this class, for example, we can mention the events of September 22, 1992 and September 8, 2002, the most lethal we will find during the period under consideration.

Class 3 (Figure 4.3 C3; six members) refers to rainfall events whose rainfall intensity is astonishingly high. Their duration is shorter than that of C2 events. Within this category was included the event of September 22, 1993, which affected the agglomerations of Marseille and Montpellier, as well as the event of December 2, 2003, which led to major material damage, especially at the level of the highway A47 [LAN 14].

Finally, Class 4 event rainfall (Figure 4.3 C4; two members) lasts longer in a larger area. Maximum intensities are lower than those of C2 and C3, but higher than those of C1. Two events are associated with this class: that of September 19, 2000, in the region of Marseille, and that of September 22, 2003, in the Gard.

The cross-analysis of severity and impact is proposed in Figure 4.4, which introduces associated impacts for each class (number of affected municipalities, number of victims, material damage expressed in millions of euros).

This first analysis shows that the resonances between hydrometeorological severity and associated impacts are not direct. In fact, we can observe a strong variability of impact within each class, while interclass differences are often not very acute.

Nevertheless, C2 seems to stand out from the other three classes, particularly when we analyze the number of affected municipalities and the number of victims. On the other hand, the impact of C2 events in terms of material damage remains almost the same as for C3.

Without providing definite conclusions, these first results highlight that the most intense events taking place on large scales of time and space are often conducive to the highest impact.

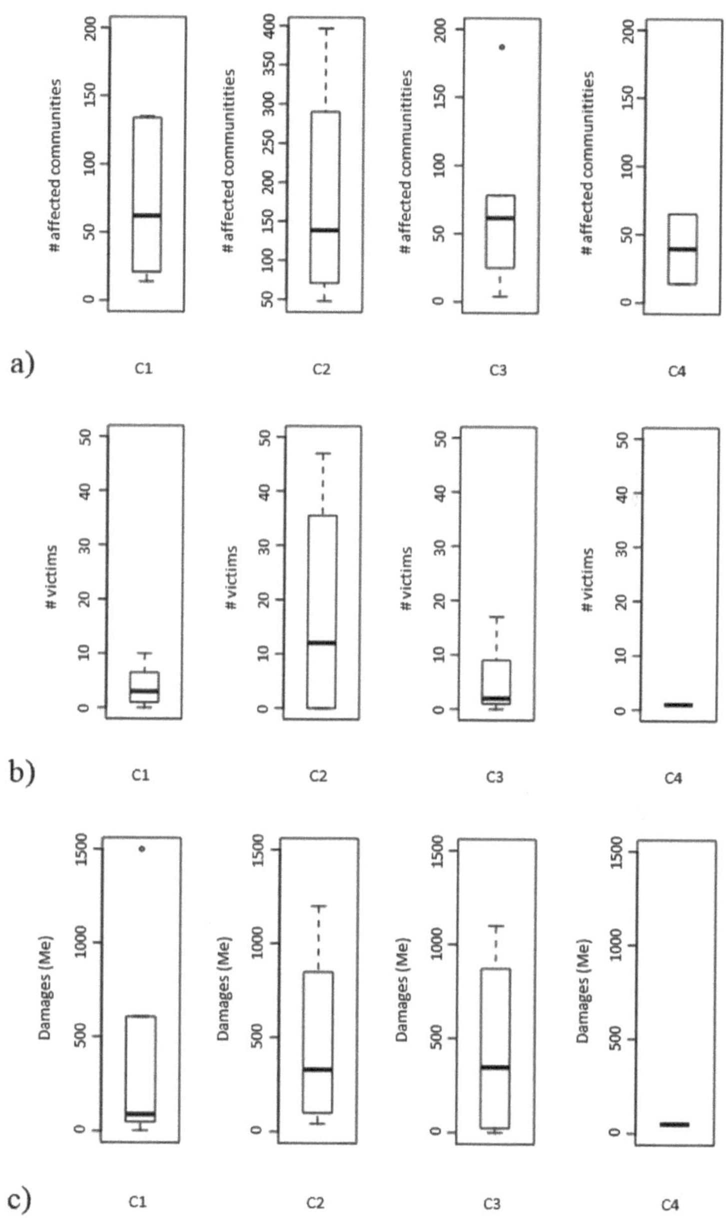

Figure 4.4. *Impact related to the flash flood events in the South East of France during the period 1988–2008 classified according to severity levels (C1, C2, C3, C4): a) number of communities affected by the events; let us observe that the presentation scale for C2 is twice the size of the other three types; b) number of deceased persons; and c) material damage in millions of euros*

While hazard severity is now well established, in virtue of its rarity and associated spatiotemporal scales, resonances between hazard and impact still remain to be ascertained. Establishing an analogy with seismic risk, which uses the Medvedev–Sponheuer–Karnik scale in order to characterize the severity of the phenomenon and its associated impact, a similar scale could be developed for assessing hydrometeorological risk. The development of this scale would require the setting up of dedicated statistical tools, built on the basis of previous events for which we have the necessary information concerning the hazard and its associated impact.

4.4. Media resonance of hydrometeorological events

This section will explore how the TV media captures flash flood events and broadcasts them. In the course of interdisciplinary work employing a space–time scale approach [GIB 00, HOL 01, CAS 06], the aim was to delineate the spatial and temporal dynamics of the physical phenomena as well as their possible imprints on media productions in order to provide a different perspective when it comes to defining the term "high-impact".

While several types of media have been explored in the framework of the CoGIS[6] interdisciplinary project (*Communication, Géosciences, Inondations, Séismes*[7]), here we will focus on describing the results obtained from the quantitative analysis of flooding and flash flood-related issues broadcast in the context of national and regional TV news over the period 1958–2012.

One of the questions concerns the scope of the different types of TV news shows in terms of the number of issues discussed (either massive or at a spatial and temporal scale). The hypothesis we suggest here is that the audience of these media varies according to their regional or national vocation, as well as their listening hours (the 20:00 TV news having a larger audience than the noon TV news show, for instance). As a result, we hypothesize that the flash flood episodes considered of lesser importance are rather reported by the regional TV news, whereas the same type of hazard with rather exceptional consequences might be accompanied by a stronger media resonance, which might influence the treatment of the subject in question, broadcast at national level during prime time. We can similarly imagine that this spatial signature might be accompanied by a temporal signature, assuming that the extent of the social crisis can be measured in terms of the persistence of the subject in the TV news over a longer or shorter period of time.

6 http://gresec.univ-grenoble-alpes.fr/version-francaise/actualites/les-archives-depuis-2011/journee-d-etudes-cogis-communication-geosciences-inondations-seismes–57327.kjsp.

7 Communication, Geosciences, Floods, Earthquakes.

4.4.1. *Media and pluviometric data*

Produced from the databases of the INA (*Institut National de l'Audiovisuel*[8]), the corpus is made up of all the TV news broadcasts produced by the three major French channels (TF1; France2; France3) over the period 1958–2012. The choice of issues integrating the corpus was decided by the results of queries using the "flood" descriptor, crossed with departmental locations (Gard, Hérault, Var and Vaucluse). On the basis of the documents collected, the number of issues and their duration (depending on the type of program) are counted by national or regional TV news, broadcast at noon or at 8 pm. The time frame studied was motivated by the availability of INA's television archives; nevertheless, we must bear in mind that the number of television channels increased during the period covered by our corpus, that is to say, the second and the third French television channels were, respectively, created in 1963 and 1973. Here, our quantitative analysis focused on the number of issues, including every type of TV news show.

In light of data concerning issue coverage and airtime devoted to the question of floods in TV news, we chose to employ daily rainfall data for the same departments used for the selection of the media corpus, that is to say the Gard, Hérault, Var and Vaucluse. Over the 1958–2012 study period, rainfall observations obtained from rain gauges located at the four target departments were extracted from the CLIMatological Data Base of Météo-France (BDCLIM). Only observations with a daily total higher than 60 mm were retained. Let us observe that since the end of the 1980s, the establishment of automatic stations improved rainfall observation processes, as well as their quality.

4.4.2. *Cross-analysis: rain intensity and rainfall extension – media resonance*

As a first approach to the period 1985–2012, Figure 4.5 presents the temporal distribution of the number of issues discussed in the TV news, in connection with the number of rain gauges, depending on whether these recorded daily cumulated rainfall lower (blue) or higher (red) than the threshold of 200 mm. The categorization of the rain according to this threshold is suggested in order to better visualize the intensity of rain; the 200 mm threshold is generally associated with extreme rainfall, for which Météo-France triggers the red warning. Thus, in Figure 4.5, the length of rain bars provides a visual cue for the spatial extent of the event. In fact, the longer the bar, the greater the number of rain gauges which recorded at

8 National Audiovisual Institute.

least 60 mm of rain. The extreme nature of rainfall is indicated by the red color of the bars, translating the number of rain gauges which recorded daily cumulated rainfall higher than 200 mm.

Figure 4.5 shows that some dates stand out not only from the point of view of rainfall extent and severity, but also from the number of TV news show broadcasts. This is the case for the events of September 1992 in Vaison-la-Romaine, episodes which followed one another from September to October 1993 on the lower Rhône Valley, September 2002 in the Gard, December 2003 on the Rhône and its tributaries and for September 2005 in the Gard. Contrary to this tendency, other events seemed to stand out in terms of media resonance, but were not so in terms of rainfall spatial and temporal extension, as was the case of Nîmes in October 1988 and Var in June 2010, for which strong intensity rainfall was registered in a small area on a single day. Finally, other rainy extended episodes, which gave rise to high rainfall cumulation, were less perceptible in the media, as were the cases of the outstandingly heavy rains in Hérault and the Rhône's flood in Camargue during October–November 1994, in Corbières in November 1999 and in the Var and the east of Rhône in November 2011.

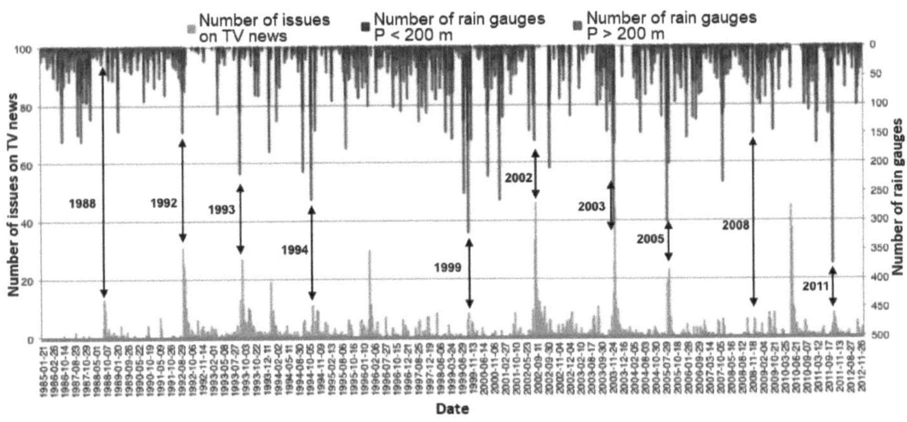

Figure 4.5. *Media resonance (orange) defined by the number of individuals affected by rain–flood episodes as reflected in regional and national TV news, as well as pluviometric hazards, characterized by the number of pluviometric stations having recorded daily rainfall discharges ranging between 60 and 200 mm (blue) and higher than 200 mm (red). Some dates of founding episodes for the context of south east France are mentioned. For a color version of this figure, see www.iste.co.uk/lutoff/hydrometeo1.zip*

On this basis, we identified three temporal frameworks where two remarkable flood events were present and for which the relationship between rainfall extension/severity and media resonance (temporality and amplitude of the response) differed. The analysis of these case studies showed that despite an identical rating of 4/6 on the severity scale used by the French official bodies (CATNAT), these rainfall events gave rise to media responses of differing importance.

Figure 4.6 shows the three analyses which crossed rainfall information (extension/rain "severity") and media resonance for the Nîmes event of October 2 and 3 1988 and that of Vaison-la-Romaine on October 21 and 22 1992 (Figure 4.6(a)), the Gard event of September 8 and 9 2002 and that of September 22 2003 in the same region (Figure 4.6(b)), as well as the Draguignan event of June 15 2010 and the one in the South of France from November 1–9 2011 (Figure 4.6(c)).

For the record, all these events were marked by very significant material damages, but also and most especially by the high number of victims (for instance, 10 and 47 victims, respectively, for the events of Nîmes and Vaison-la-Romaine; 24 and 1, respectively, for the events of September 2002 and September 2003 in Gard, 23 and 0, respectively, for the episode of Draguignan in June 2010 and November 2011 in the south of France).

Overall in Figure 4.6, we can observe that media response time ranged between 1 and 2 days after very intense rainfall began to be observed. Therefore, this response time indicates that communicated information was mainly focused on a description of the event, its associated consequences and was not enough to offer useful information on how to get protected against the ongoing event. Other media, such as social media (for example, Twitter) or YouTube are probably more appropriate for these purposes, but we shall not analyze them here. Let us equally observe that the signature of meteorological forecasts and vigilance alerts communicated in TV news have not been specifically explored in this study.

Figure 4.6 shows that there has been an increase in the number of flood-event related issues over time. While this is less true in Figure 4.6(c), where the last (non-lethal) event produced fewer issues than its predecessor, Figures 4.6(a) and (b) show trends where the second event triggered the highest amount of issues. By analyzing in detail the issues discussed in TV news, it appears that the second event echoes the first event, in such a way that the issues covered concern both the current event as well as the nearest past event, as a kind of reminder of their consequences, the measures taken, etc. Reminders of previous highly deadly events have also been the subjects of many issues covered by TV news when ongoing rain is observed. This is the case, for example, in Figure 4.6(b), where several issues discussed on the news during the events of October 7 and December 10 2002 behaved as "reminders" of the events of September 2002; let us also observe that the events of October and December 2002 did not cause major flooding. In this context, quantifying an event's media resonance is not direct and requires a precise textual analysis of the issues discussed.

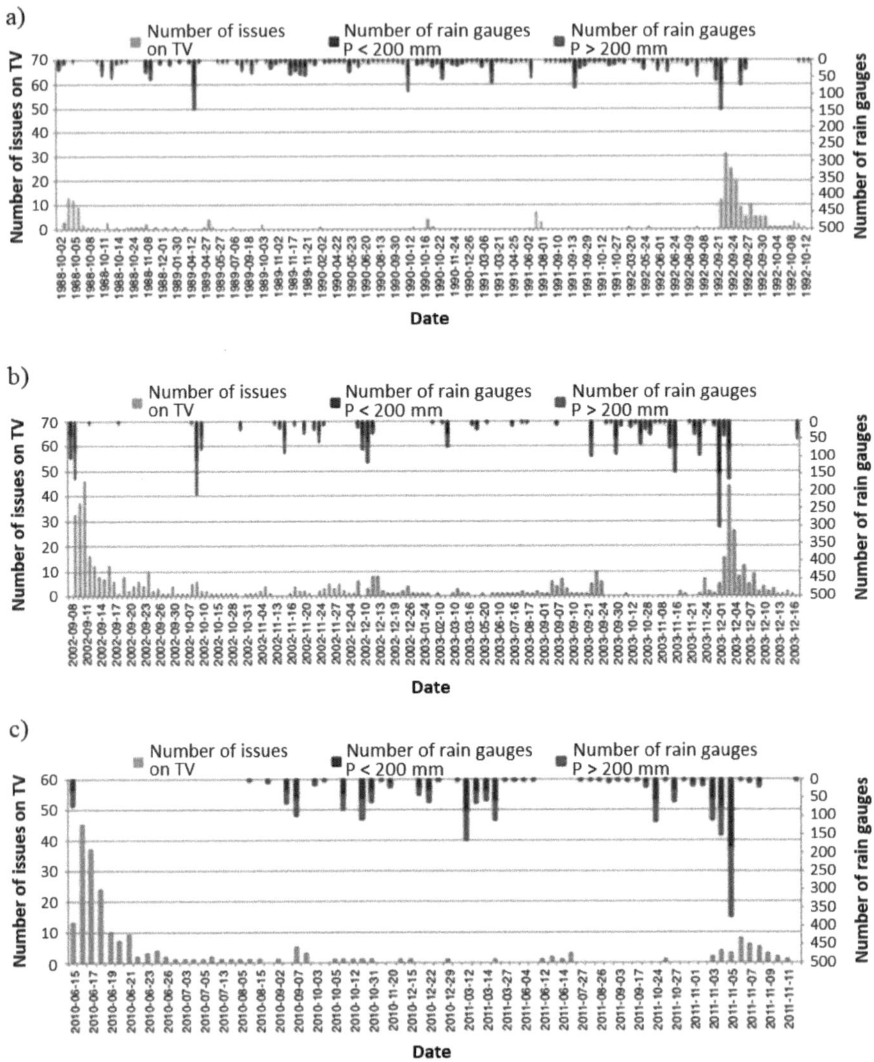

Figure 4.6. *Media resonance (in orange, the number of individuals affected by rain–flood episodes as reflected by regional and national TV news) and pluviometric hazards (number of pluviometric stations with a daily rainfall discharge ranging between 60 and 200 mm (blue) and higher than 200 mm (red). a) Nîmes events of October 2 and 3 1988 and from Vaison-la-Romaine, from October 21 and 22 1992; b) Gard events of September 8 and 9 2002 and September 22 2003; c) Draguignan events of June 15 2010 and from the south of France from November 1–9 2011. For a color version of this figure, see www.iste.co.uk/lutoff/hydrometeo1.zip*

Establishing a connection between the number of issues discussed and the "severity" of the event is therefore not immediate. Figure 4.6 suggests that the greater the extent and the intensity of the event, the greater the number of issues discussed. However, the episode of Draguignan in June 2010 (Figure 4.6(c)) was the trigger for a number of issues discussed on the TV, almost identical to those of the episode of September 2002 (Figure 4.6(b)), but the spatial extension and intensity of the Draguignan event were far smaller.

4.5. Conclusion

In this chapter, several leads were explored so as to identify specific elements that could help us to define a high-impact weather event. The simple definition including hydrometeorological characteristics (for example intensity, duration, extension of the pluviometric hazard and associated flows) is not enough, even if this first definition could help us to implement alerts, and thus contribute to mitigating the hazard's impact. Characterizing a high-impact weather event requires being able to establish links between the hazard and the human and material consequences suffered. Hence, it is necessary to have specially dedicated databases including information about the hazard and the impact, expressed by means of data regarding not only material and human losses but also data concerning the event's resonance (that is to say, how its effects expand, by means of media and, nowadays, social networks as well). Indeed, it might be interesting to understand the evolution of this acceptability over time in order to refine the concept of population vulnerability to these hazards. Finally, on this basis, the development of a "severity" scale, as has been developed for seismic risk, is a necessary objective to be attained, thoroughly based on such data.

4.6. Bibliography

[BEN 16] BENFIELD A., Annual global climate and catastrophe report, AON Empower Results, Risk Reinsurance Human Resources, 2016.

[BOI 13] BOISSIER L., La mortalité liée aux crues torrentielles dans le Sud de la France: une approche de la vulnérabilité humaine face à l'inondation, PhD Thesis, Paul Valéry University, Montpellier III, 2013.

[BOU 15] BOUDOU M., Approche multidisciplinaire pour la caractérisation d'inondations remarquables. Enseignements tirés de neuf événements en France (1910–2010), PhD Thesis, Paul Valéry University, Montpellier III, 2015.

[BOU 17] BOUDOU M., LANG M., VINET F. *et al.*, "Comparative hazard analysis of processes leading to remarkable flash floods (France, 1930-1999)", *Journal of Hydrology*, vol. 541, pp. 533–552, 2017.

[CAS 06] CASH D.W., ADGER W., BERKES F. *et al.*, "Scale and cross-scale dynamics: governance and information in a multilevel world", *Ecology and Society*, vol. 11, no. 2, pp. 8, 2006.

[CER 11] CERESETTI D., ANQUETIN S., MOLINIÉ G., *et al.*, "Severity diagrams: a new approach for the multi-scale evaluation of extreme rainfall events", *Weather and Forecasting*, vol. 27, pp. 174–188, 2011.

[DRO 14] DROBINSKI P., DUCROCQ V., ALPERT P. *et al.*, "HyMeX: a 10-year multidisciplinary program on the Mediterranean water cycle", *Bulletin of the American Meteorological Society*, vol. 95, pp. 1063–1082, 2014.

[GIB 00] GIBSON C.C., OSTROM E., AHN T.K., "The concept of scale and the human dimensions of global change: a survey", *Ecological Economics*, vol. 32, pp. 217–239, 2000.

[GUH 13] GUHA-SAPIR D., HOYOIS P., BELOW R., Annual disaster statistical review 2012: The numbers and trends, Centre for Research on the Epidemiology of Disasters (CRED), Institute of Health and Society (IRSS) and the Université catholique de Louvain, Brussels, 2013.

[GRA 14] GRANDA S., Critique et analyse des bases de données sur les catastrophes naturelles pour la caractérisation des événements hydro-météorologiques à forts impacts socio-économiques, Master's Thesis, Joseph Fourier University, Grenoble, 2014.

[HOL 01] HOLLING C.S., "Understanding the complexity of economic, ecological, and social systems", *Ecosystems*, vol. 4, pp. 390–405, 2001.

[IPC 14] IPCC, "Summary for policymakers", in FIELD C.B., BARROS V.R., DOKKEN D.J. *et al.* (eds), *Climate Change 2014: Impacts, Adaptation, and Vulnerability. Part A: Global and Sectoral Aspects. Contribution of Working Group II to the Fifth Assessment Report of the Intergovernmental Panel on Climate Change*, Cambridge University Press, Cambridge, 2014.

[JON 05] JONKMAN S.N., "Global perspectives on loss of human life caused by floods", *Natural Hazards*, vol. 34, pp. 151–175, 2005.

[KHO 16] KHODAYAR S., FOSSER G., BERTHOU S. *et al.*, "A seamless weather–climate multi-model intercomparison on the representation of a high-impact weather event in the western Mediterranean: HyMeX IOP12", *Quarterly Journal of the Royal Meteorological Society*, vol. 142, pp. 433–452, 2016.

[LAN 14] LANG M., CŒUR D., *Les inondations remarquables en France, Inventaire 2011 pour la Directive Inondation*, Edition Quae, Paris, 2014.

[LAN 16] LANG M., COEUR D., AUDOUARD M. *et al.*, "BDHI: a French national database on historical floods", *3rd European Conference on Flood Risk Management*, Lyon, France, October 2016.

[LLA 13] LLASAT M.C., LLASAT-BOTIJA M., PETRUCCI O. *et al.*, "Towards a database on societal impact of Mediterranean floods within the framework of the HYMEX project", *Natural Hazards Earth System Science*, vol. 13, pp. 1337–1350, 2013.

[OEC 14] OECD, Étude de l'OCDE sur la gestion des risques d'inondation: la Seine en Île-de-France 2014, OECD Publishing, Paris, 2014.

[OLS 17] OLSSON J., PERS B.C., BENGTSSON L. *et al.*, "Distance-dependent depth-duration analysis in high-resolution hydrometeorological ensemble forecasting: A case study in Malmo City, Sweden", *Environment Modeling & Software*, vol. 93, pp. 381–397, 2017.

[PAP 13] PAPAGIANNAKI K., LAGOUVARDOS K., KOTRONI V., "A database of high-impact weather events in Greece: a descriptive impact analysis for the period 2001–2011", *Natural Hazards Earth System Science*, vol. 13, pp. 727–736, 2013.

[RAM 05] RAMOS M.H., CREUTIN J.-D., LEBLOIS E., "Visualization of storm severity", *Journal of Hydrology*, vol. 315, pp. 295–307, 2005.

[SHA 17] SHABOU S., RUIN I., LUTOFF C. *et al.*, "MobRISK: a model for assessing the exposure of road users to flash flood events", *Natural Hazards Earth System Science*, vol. 17, pp. 1631–1651, 2017.

[TER 17] TERTI G., RUIN I., GOURLEY J.J. *et al.*, "Towards Probabilistic Prediction of Flash Flood Human Impacts", *Risk Analysis*, October, 2017.

5

Legal Evolution of the Conflict Between Water Mobility and the Mobility of People

5.1. Introduction

At an institutional level, avoiding the destructive impact of floods by reducing population exposure to hazards demands following a twofold approach: first, it requires constraining water mobility via the construction of protective or storing structures (*active protection*) and second, it implies regulating not only the mobility of people, but also of property, depending on the territory's flood ability (*passive protection*)[1].

As contemplated in this book, the mobility of people in the face of flooding refers both to the individual who determines their position at a certain moment (daily mobility), as well as their habitat, which not only refers to the position of their home, but also their working premises or leisure facilities (residential mobility)[2]. In fact, the type of mobility discussed in this chapter is essentially residential. Daily mobility involves the undisputed principle of freedom to come and go as well as road regulations, but to our knowledge, it does not suggest binding legal provisions

Chapter written by Jean-Dominique CREUTIN and Valérie SANSEVERINO-GODFRIN.

1 The opposition between passive and active protection is widely spread in technical texts dealing with natural disasters (see, for example, *Les mots des risques naturels*, J.M. Antoine, Presses Universitaires du Mirail, Toulouse, 2008).

2 The opposition between daily and residential mobility corresponds to that of short-term and long-term mobilities, as described in Chapters 2 and 8.

limiting the freedom to circulate[3], in the way of regulations concerning the transport of dangerous substances.

The purpose of our reflection in this chapter mainly focuses on the institutional response given by the authorities over the past two centuries in view of reducing flood exposure.

In this chapter, we will refer to the three "pillars" defined by W.R. Scott [quoted in HOF 99], so as to characterize the institutional perspective, making a distinction between:

1) the *regulatory* (or legal) institution that rules organizations through laws and regulations and their enforcement, for example, in the way that municipalities reduce the risk exposure of their inhabitants by applying the guidelines of a Flood Risk Prevention Plan, without which they would become exposed to government sanctions;

2) the *normative* (or social) institution, which comprises a certain operational know-how, together with the teaching of accompanying methods that organizations use for protective or preventive purposes (alerts), and finally;

3) the *cognitive* (or cultural) institution, which relates to a certain way of thinking about floods, for example, through the words used to describe them and to express understanding.

This chapter will focus on the *regulatory* institution, but it is obvious that *normative* and *cognitive* aspects underpin legal evolution, which encompasses and structures our analysis. The chapter is organized into three sections.

In the first section, we will attempt to identify two major trends in how the institutional response to flooding evolved in France throughout the past three centuries. The first tendency transformed flood protection from being an individual concern to becoming a public safety issue, while the second one arose from a desire to control water movement and from a need for regulating residential mobility. These two trends are not directly related, but both feature a search for balance between (1) the respect for private individual property – since exposure control means servitude and restriction of the right to enjoy one's own property and (2) the social and territorial solidarity that flooding entails by its very nature – exposure control also means concertation and a collective planning of space.

3 Circumscribed regulations in time and space (prefectorial or municipal orders) may restrain circulation liberty, depending on events. On the other hand, meteorological vigilance bulletins published by *Météo-France* respond to a different type of institutional action, with no obligation for people to respect suggested recommendations, thus without any legal effect.

The second section will focus on the multiscale nature of flooding and institutional responses to it. High-impact hydrometeorological events can refer to scales widely ranging from large rivers flooding major cities, whose paralysis threatens the national economy, to small rivers triggering flash-floods, often resulting in casualties. Regulatory institutions also interlock legal and territorial levels, ranging from national and European laws and directives to their applications by local collectivities. While these two "hierarchies" are not directly linked[4], they do entail a proper application of the subsidiarity[5] principle with regard to the public action of flood management.

In conclusion, the third part will summarize the thread of guiding ideas concerning institutional responses to flooding. It will emphasize the implications of normative and cognitive institutions in the major evolutionary stages described by suggesting their classification around four founding questions.

All things considered, the itinerary described below reveals the French institutional construction of a "collective" conception of flooding, as it was modeled from the 17th Century until recently. This construction is essentially based on an analysis of the complicated interactions between water and urban construction. It required centralizing thorough knowledge of the functioning of rivers in order to distinguish the effects of natural and anthropic elements on this interface. Finding balance reflects a socioeconomic trade-off between planning constraints in ordinary times and the need for solidarity at the time of disasters.

5.2. A swing between waterstream constraints and human settling constraints

It is the anthropization of natural spaces near rivers or along the coastline, and their economic use, that raises the question about flood protection – rare and unexpected invasions of these spaces by water, putting people and property at risk. What should be done in order to avoid the consequences of floods? Should homes and activities be protected, by sheltering them with protective devices? Should the natural mobility of watercourses be preserved, by limiting the extension of human settlements? Is this really an alternative or are these complementary actions?

In the following, we will show that the institutional response considers these as complementary actions, whose balance has considerably shifted over the last two centuries. As we can see in Table 5.1, by introducing the legal texts which have

4 Each legal level presents interest to floods occuring at all scales – from minor tributaries to major rivers.

5 The subsidiarity principle considers that the competent entity closest to those directly concerned by a public action must be accountable for it.

concerned water management since the 17th Century, the idea of preserving the liberty to occupy space and to protect oneself individually is slowly giving way to public safety and urban planning regulation. It was only at the end of the 20th Century that the two mobilities were subjected to special regulation, making sure to draft texts that would ensure consistency between the two issues.

Date	*Laws and founding events*		*Postevent evolution*	
1669	Royal decree on waters and forests	Law on free water mobility and its protection by the public force (Royalty in this case). Did not address the question of flooding.	"Hollow" appearance of the flooding problem in the law, stemming from water-related questions (1669–1858).	
1807	Law concerning swamp drying-out	Law on property rights over dried-out swamps. It attributed the responsibility of control over water mobility to the owner who benefited from it. For the first time, it addressed the question of protection against water invasion.		
1854	*Rhône, Loire and Garonne floods*			
1856	Law on drainage	Law on drainage incentives. As the law was passed under the influence of disastrous floodings, it led to a plea for passive protection against flooding through a slowdown of floods.		
1858	Law on city defense work against flooding	Law authorizing the administration to order dike building, transfer or destruction. Setting of Flood-prone Area Plans.	Law orientation toward flood containment and flood slowdown (1858–1935) – technical vision.	

1860	Law concerning mountain reforestation	Law encouraging mountain reforestation. No clear reference to flooding.		
1898	Law on water regimes	It passed regulation on water ownership rights, but did not address the question of flooding.		
1910	*Seine and Loire floods*			
1910	Law for flood-related expenses	Law enabling temporary credit after the Paris 1910 floods. Not broadly extensive toward the question of flooding.		
1914	Bougère law project	Advocated for flood slowdown, plea for private damage financing, pointed out planning mistakes.		
1930	*Garonne and Tarn floods*			
1935	Decree-law on waterflow	Law regarding residential mobility for the first time, by setting public use precautions and limiting construction through the delimitation of submersive surface maps (SSM).	Emergence of control over residential mobility (1935–2007) – anthropocentric vision.	
1955	Decree for applying article 91 of the French Building Code	Decree enlisting the reasons for rejecting construction permits. Subordinated building activities on terrains prone to natural disasters, (including flood risks), to special conditions.		

1964	Law regarding water regimes and water distribution, as well as the struggle against pollution	Regulated the struggle against pollution as well as water regimes and water distribution (waterstreams, subterranean water).		Legal acknowledgement of the multiscale character of floods as well as the multiple uses of water (1964–2016).
1982	Law regarding the compensation of victims of natural disasters	Law defining the role of insurance in the occurrence of "natural disasters", implementing the Risk Exposure Plan (REP), its regulation and its maps.		
1992	Water law	Defined the role of the REP in the conservation, restoration and influence of flooded plains.		
1993	Decree regulating the delimitation of areas exposed to natural risk	Decree outlining the mapping of REPs and the conservation condition of flooded plains.		
1995	Law regarding the reinforcement of the environment's protection	Law on the participation of the general public in environmental matters, the prevention of natural risk catastrophes, REPs, management of green spaces, waste management and pollution prevention.		
1999	*Aude, Hérault and Tarn floods. Lothar and Martin storms*			
2001	*Somme flood*			
2002	*Gard flood*			

2003	Law regarding risk prevention… and damage repair	Made it possible to establish precautions for creating retention areas and "water mobility areas".		
2006	Law regarding water and water environment			
2007	Directive regarding flood risk assessment and flood risk management	Directive calling for flood risk assessment, its cartography and the elaboration of a management plan against this type of risk in concert with the whole of Europe.		
2015	Law regulating the new organization of the French territory	Law transferring the exclusive and mandatory competence for managing water environment and for preventing flood risk at an intercommunality level (GEMAPI).		

Table 5.1. *Summary of the different legal texts regarding water mobility and mobility of people, list of founding flood events*

5.2.1. *An implied appearance of the flooding problem in legal documents (1669–1858)*

A law regarding the concern for the free flow of water, the Royal Ordinance of August 1669 responded to Louis XIV's request to "remedy the disorders which had affected the administration of water and forests of the kingdom[6]". It included the

6 The order's preamble stated: "Despite the fact that the disorder which slipped in the waters and forests of our kingdom was so universal and incorrigible that the remedy seemed almost impossible; however, the heavens have blessed us with the application of eight years which we have devoted to the recuperation of this noble and precious part of our domain, that we see it today flower anew more than ever, and to produce with such an abundance for the public that all kinds of advantages can be expected, either for private life privileges, for warfare

entirety of royal and seigneurial rights, as well as competent jurisdictions over the use of woods and rivers, as well as fishing and hunting, ordered under a single and coherent framework[7]. It dealt with all the "enterprises and claims on navigable and buoyant rivers[8]", under Paragraph 27, from which the rest of the quoted articles are extracted. When considered in conjunction with "highways[9]", the river is seen as a transportation system whose operation must not be interfered with, neither by energy, materials nor timber production activities. In fact, Article 42 states:

> "Nobody [...] shall be able to build mills, cofferdams, floodgates, fisheries, regulation straits, walls, tree plans, stone piles, earth or roots, nor any other harmful buildings or hindrances in the course of water in rivers and navigable and buoyant rivers, nor throw in any kind of trash [...]". Article 43 contemplates obstacle demolition and Article 44 bans "everyone from diverting water from rivers [...] or from weakening and altering the course by trenches, ditches and canals".

The ordinance did not properly address the issue of flooding, despite the fact that free water mobility (which it confirmed and placed under the protection of the public authorities) was one of its key aspects.

The drying up of swamps, that is to say removing land from the permanent invasion of water, was a new opportunity to legislate in relation to rivers. The law of September 16, 1807 essentially addressed the economic consequences of drainage as announced in Article 1: "The ownership of swamps is subject to special regulations". Among these regulations, we can mention the "conservation of drying works" (Paragraph 6), leading to the question of dikes as an urban planning element among others (Paragraph 7: "navigation works, roads, bridges, streets, squares and quays of cities, dikes, sanitary works of municipalities"). Without directly dealing with flooding, the law of 1807 confirmed the principle of the authorization for work construction in the beds of watercourses (Article 27: "The conservation of drying

needs or finally, for the coronation of peace and the multiplication of commerce, thanks to long distance voyages in all directions of the world".

7 This global approach makes us think with our contemporary analysis grid toward an environmental law; at least, it points to the need for a public force intervention in the arbitration of different water uses.

8 Formula extracted from Paragraph 1 Article 3. Buoyant watercourses are those rivers in which wood can be transported by letting it float, without the need to explicitly "drive" it.

9 Here, we interestingly encounter the proximity between water mobility and people mobility.

works, that of dikes against torrents, streams, rivers and on the shores of lakes and the sea, is ruled by the public administration"). In Article 33, the law clearly states the principle (which is still in force nowadays[10]) that riparian owners must ensure, or at least contribute, to the protection against the natural action of water:

> "When it comes to building dikes at sea or against navigable or non-navigable rivers, streams and torrents, the need will be confirmed by the Government, and the expense borne by protected property, in proportion to their interest in the works; except in those cases where the Government might consider it to be useful and fair to provide relief with public funds".

The economic questions related to land drainage remained significant during the 19th Century, as it was shown by the law on drainage of July 17–23, 1856[11]. While the term flood was not explicitly mentioned in the law of 1807, the government nonetheless became aware of the shortcomings of the new Civil Code in terms of water regulation, as early as that time. In virtue of the decree from May 19, 1808, it opened the discussion about a draft rural code dealing with the soil regime, the water regime and rural policy, in 3 volumes; a policy that would still take a century to be implemented [SIR 98].

In this implied appearance of floods in legal documents arising from water-related issues, a change in point of view had taken place between the Royal ordinance of 1669, and the law of 1807: we moved from the question of guaranteeing the free flow of water to that of protecting population against its eroding action[12].

---

10 The responsibility of private owners may be elicited in the absence of protection works, for example, on the part of trade union associations, who may have avoided dredging or containment works necessary for water evacuation (Administrative Court of Appeal of Lyon, March 1991, no. 89LY00968).

11 The law regulated State loans inviting land drainage.

12 This change is contemporary to what certain authors have considered as the "fusion" between science and engineering. This fusion is imprinted with a Cartesian conception of nature on the one hand (let us recall that according to R. Descartes, in his *Discourse on the Method*, human beings are assimilable to "masters and possessors of Nature") and, on the other hand, the ideology particularly set in motion since the Enlightement century regarding the capacities of science and technology to master nature and its manifestations. This fusion began throughout the 17th Century and became a universal phenomenon by the end of the 18th Century. "By this time, modern industry as well as the science it presupposes began to achieve the dream of domination over nature expressed by Descartes" – see Fontaine [FON 08].

5.2.2. *Legal orientation toward flood containment and flood slowdown (1858–1935) – a technical vision*

The law of May 28, 1858 responded to the great floods of the mid-19th Century, and particularly to those of the Rhône, the Loire and the Garonne in 1856, all of which caused considerable damage. These floods pushed the State to become involved in "the defense of the cities and the defense of the valleys[13]" and the law[14], in its first article, declared: "The State will proceed to the execution of the works intended to protect the cities from flooding". For the first time, there was a law dealing with flooding and the suggested response was one of "active" protection containment works[15]. For over a century, this law focused public action on the constraint of water mobility, while allowing for the free mobility of people. It reflected an approach to the prevention of natural risks[16] that would eventually become dominant with the development of "Grand Corps of the State[17]".

Concerning the defense of the cities, the studies adopted by the Civil Engineering General Council (*Conseil général des ponts et chaussées*) revealed a certain technical enthusiasm:

> "The defense of the cities does not present very serious difficulties. In most cases, it is a matter of consolidating the current dikes and raising them above the level of the largest floods known to date" (Report by Member of the Louvet Parliament for the presentation of the law of May 28, 1858, in [CHA 61]).

13 Letter from Napoléon III to his minister responsible for Public Works, Agriculture and Commerce, dated from July 19, 1856, quoted as one of the motives for the law of 1858.

14 The text for this law can be found at: http://gallica.bnf.fr/ark:/12148/cb375780597/date.

15 Let us quote here a phrase extracted from a speech by Napoléon III: "I honourably declare that in France, rivers, as well as revolutions, go back to their beds without being able to come out again", which illustrates our assumption regarding water mobility, providing a more personal viewpoint in what concerns people mobility.

16 Different authors have referred to a "technocentric paradigm" with regard to this approach. For water matters, it is typical of "engineering culture" and "privileges protection" [BEU 08] in a debate opposing "technocentrism and environmentalism", or "developers vs. ecologists" [ROD 10]. This name, inherited from sociology of work [LOJ 82], can be argued; in the measure that it completely excludes a positive alliance between engineering and the environment.

17 In particular, forestry engineers (see the laws of July 28, 1860 on mountain reforestation and from April 4, 1882 on mountain land restoration) and military engineers, who were at the origin of the Corp of Civil Engineers.

Regarding "rural floods [which] manifest in much larger proportions" than those of cities, the question was immediately postponed, partly for economic reasons[18], but also probably because of the awareness about the limits of an exclusively "active" protection approach.

Dikes protected property located alongside the dikes themselves, but could also change the flow of water during floods and even aggravate the effects of flooding[19]. Louvet developed the interesting notion of "offensive" works, in the sense that they could create problems elsewhere:

> "If the State justifiably refuses to allow or maintain offensive dikes in the valleys, it would be unfortunate if the works planned for the defense of the towns were themselves offensive to the neighboring countryside. This idea has worried your committee[20]" [CHA 61].

The studies carried out in preparation for the 1858 law reflected the hesitation of engineers on the type of intervention to be favored. As a rapporteur, Louvet evoked the "passive" alternative of flood reduction:

> "Now, apart from several other delicate questions, we immediately find ourselves faced with a first problem that is very difficult to solve: shall we limit ourselves to raising the dikes, or shall we seek a way to lower flood levels, either by reforesting mountain peaks and sides, or by clearing the bed of our streams and rivers of obstacles that hinder

18 See Louvet's observation, infra. Also see Guerrin [GUE 14] who explained that 19th Century policies were not translated into controlled floodings of rural areas near big cities because these areas were often very fertile from an agricultural point of view.

19 In fact, dikes may cause waterstream retention or acceleration that may provoke water increments upstream or downstream. Designed for a local protection, they globally disturb the flood field, thus turning a privated initiative into a public problem, hence the need for controlling their implantation.

20 A similar comment was made by the president of the legislative body regarding the flood of 1856 at Lyon: "Lyon, in the distribution of State aids, is attributed with around 13 million francs, an undoubtedly considerable sum, but naturally justified, if we bear in mind that during the flood of 1856, 1,200 houses were partially or completely destroyed. The reparation and heightening of the Rhône and Saône quays defend the Empire's second important city against accidents, a fact which contradicts an advanced civilization as ours. Have we not seen, barely a few years ago, certain neighbourhoods, certain Parisian quays, that of the *Hôtel de Ville*, for example, submerged and furrowed by ships instead of carriages? The aim is to preserve Lyon, in the same way that Paris has already been preserved". The reference to Paris presumably points to the flood of February 8, 1850, whose height at the Austerlitz Bridge was 2 m lower than that of 1910, with the accompanying results we know [GAS 54].

> the rapid flow of water, or shall we even create new beds where current opportunities are recognized as insufficient, by establishing artificial reservoirs intended to retain the water coming from the great rains or the snow melt in the upper part of the rivers and main reservoirs, and thus prevent the sudden irruption of these waters in the plains?" [CHA 61]

It is interesting to observe that in the months following the catastrophic floods, the law of July 17–23, 1856 was passed, concerning drainage assistance which, apart from confirming the State's will to promote "wetland draining", led its advocates to assert that "drainage, applied to many of the millions of hectares that dominate the valleys of our most formidable rivers, is one of the most powerful means of warding off the instantaneity that causes flood danger, since it delays the arrival of a portion of the tributaries[21]". In the years that followed, the debate on the role of forests on flood formation was intense and often very difficult, due to a lack of observation means[22]. The law of July 18, 1860 on mountain reforestation mainly responded to economic motivations, rather than hydrological ones.

In the end, the law of 1858 established a local compromise between the mobility of watercourses and property protection. In fact, it aimed to ensure the free flow of water, authorizing the administration to build, move or even remove dikes:

> "Any dike established in the valleys [...] and proved to hinder the flow of water or to restrict flood plains in a harmful way, may be moved, modified or removed by order of the administration, except, if applicable, for the payment of damage compensation" (Article 7 of the 1858 Act) [TRO 47].

On one hand, the implementation of the *Plans de Zones Inondées* (PZI) (Flooded Area Plans) enabled the local control of dike settings. On the other hand, the law aimed to build protections for the main cities of the French territory located along the Rhône, the Loire, the Garonne and the Seine (together with 13 tributaries).

But the 1858 law focused on the management of dikes for large rivers' urban areas. It temporarily left aside the difficult question of rural valley flooding *de facto*, as well as that of other elements that could aggravate flooding[23]. It was thanks to the draft rural code initiated in 1808, relaunched several times in 1818, 1834 and 1854 and finally filed by the government in the Senate on July 13, 1876, that the

21 Laws, decrees and legislative acts, 8th and 9th books, 4th section, 1856, p. 104.

22 As an example, see a controversial viewpoint [VAL 59] or rather a careful technical description of the state of the art in those times [MON 58].

23 In particular, in 1876, the Lorraine valleys posed the question around the role of Mine Waste Dump and the Loire, that of trees, both slowing down and log jam factors.

successive public authorities tried to complete the construction of a legal framework for flood prevention. Transmitted to the House on November 20, 1883, it went through several successive commissions, the long progress of Book 2 of the rural code project was achieved by the law of April 10, 1898, regulating water regimes and fixing the right of water property ownership, but it no longer dealt with the question of flooding (General Compendium of Laws, 1898). The 1860–1900 period marked the collapse of the dynamics created by the floods of 1856 and even an undeniable retreat from the problem – the rather lukewarm law of 1858, arising from the problem of major valley cities and limited to dike planning, was faced against individual property considerations brought forward in the rural code's section dedicated to water. Flooded area plans were neither elaborated (with the exception of the Rhône River in 1911) nor applied.

The Paris flood of January 1910[24] (the highest waters took place on the 29th) was sufficiently shocking at the national level so as to bring about an immediate and outstanding response. The A. Picard commission, set up on February 9, examined the causes of the disaster in a series of reports covering a wide range of points and making a myriad of proposals[25]. Compensation for damage was the subject of a government bill "tending to provide the Minister of the Interior with provisional funds for flood-related expenses during the year 1910", as from March 22, or less than two months after the flood peak. Via a ministerial dispatch of December 21, 1911, the State invited the Paris City Council and the Departmental Council of the Seine to undertake a major recalibration work on the Seine and its tributaries, with its support [LEM 12][26]. On July 14, 1914, on the eve of the war, a deputy tabled a bill "for the establishment and publication of reports for every river in the Loire basin, with the aim of assisting the fight against flooding". At the same time, this proposal recalled that the Loire had also been severely affected in 1910, but without receiving the same attentions as Paris, and that it is necessary to improve the

24 This Seine flood was accompanied by a similarly important Loire flood, but the "economic" impact of the Parisian paralysis caught the public's attention, and particularly that of the legislators.

25 The Commission structured its work around 10 topics, including (1) the extent and height of the flood plain – Boreux-Tur plan, Alexandre map of Paris surroundings, Nouailhac-Pioch flood monography; (2) flood forecasting (its precision, its broadcast) – reinforcing the service opened in 1854 by Belgrand, particularly concerning observation and broadcasting means; (3) the condition of protection works – heightening and regularizing quay gaps; (4) public hygiene: making drains hermetic and energy production units unsinkable; (5) internal and external communications of the capital city: roads, railways – protecting and draining railway lines, but also telephone and telegraph ones – isolating water cables; (6) gas and electricity distribution; (7) slowing down upstream floods.

26 The Paris Council response referred to (1) work rehabilitation and immediate measures; (2) measures destined to the protection of the city of Paris and its surroundings; (3) great flood forecasting works.

knowledge of the watercourse regime[27]. It pointed out the multiple planning errors which had led to the excessive shrinkage of the river beds[28].

The 1914–1918 war period undoubtedly put an end to the worries generated by the Paris flood of 1910, and probably also delayed the evolution of flood prevention[29]. For example, it is interesting to observe that the idea of flood forecasting, already posed after the floods of 1856, was again carefully considered by the Picard commission, revealing a desire to influence the mobility of people at the time of the crisis – *daily mobility*.

5.2.3. *Emergence of residential mobility control in the law (1935–2007) – toward an anthropocentric vision*

It was not until 1935 that the question of water mobility and of the protection of people and property was raised again, as a consequence of the catastrophic floods caused by the Garonne and the Tarn in the spring of 1930. The legislative decree of October 30, 1935 "on the measures to be taken in order to ensure the flow of water[30]" and its public administration regulation of October 20, 1937 resumed the national institutional effort begun in 1858 so as to regulate the submersible parts of the valleys. Because of its limitation to dikes alone, and its restricted application to a reach of the Seine and the Rhône, the law of 1858 had left "the administration [without] any weapon enabling it to limit, as little as possible, the freedom that local reparians enjoyed for using their land[31]". The decree law of 1935 marked an important turning point in terms of population protection because it explicitly mentioned, probably for the first time, a mastery effect of residential mobility

27 Law proposal by F. Bougère, Maine and Loire local representative, whose report extensively developed his viewpoint on passive protection measures for flood slowdown, and asks that damage count be extended to private property, and also pointed out the development mistakes that complicate active protection.

28 "Has not one reach of the Maine been filled at the town of Angers, which would later create on this site a vast fairground and now complains about the invasion of neighboring great waters? – Tell me about any riparian city where at least one district has not been entirely flooded in its valley? – Or has not the State accepted and directed works of this kind?"

29 It is nonetheless remarkable that in the middle of the most critical period of the war, the Chamber of Deputies continued on January 25, 1917 to vote in favor of credits for works recommended by the A. Picard Committee (*Report making a plea for the vote of a credit of 1,000,000 francs in concept of the State's contribution to the preparation of a preliminary draft for deepening the Seine between Port-à-l'Anglais and Rouen*, by A. Louppe, local representative at the Chamber of Deputies, no. 2931, 11th Parliament, Session 1917).

30 Public administration decree of October 20, 1937, Communication no. 34 of April 5, 1952.

31 Minister's report to the President of the Republic, presenting the decree of October 30, 1935 – JORF p. 11688.

by controlling constructions in flood-prone areas, and not only dikes. Article 1 stated that:

> "subject to the provisions of this decree are the establishment or maintenance of dikes, embankments, cumbersome deposits, fences, plantations, constructions or any other works likely to hinder the water flow or to restrict the flood plains on the submersible parts of designated watercourses in a detrimental manner".

In the explanatory memorandum, the President of the Council observed that:

> "As regards to buildings, such a regulation would offer a twofold advantage: on the one hand, it would have the effect of protecting builders from the certain risks they might incur when building in a dangerous area, and on the other hand, it might prevent increasing the burden that the Treasury bears considering that the State is necessarily committed to coming to the aid of flood victims".

Article 2 introduced a new tool, the Submersible Surface Plan (SSP), designed to identify flood-prone areas and thus be able to regulate any actions or works in these areas, which might be detrimental to the conservation of floodplains. At the same time, the aim of the SSPs was to establish public utility arrangements and to maintain the free flow of water by limiting building development in flood-prone areas.

The government[32] – or the engineers at the ministries at least – were simultaneously certain about "the usefulness of a general regulation for the conservation of floodplains" and puzzled by the complexity of the very notion of "conservation", which involved taking advantage of two contradictory actions: flood capping by storing water in floodplains and flood limitation through containment. The idea of leaving a place for the river, of "keeping intact the role that nature has assigned to valley floodplains as flood regulators" was supposed to be articulated with ongoing and future planning in flood-prone areas, which was not limited to a matter of containment, as the law of 1858 had envisioned. Even if the system mainly concerned the most important national watercourses listed in Article 48 of the *Code du domaine public fluvial* (Public Rivers Code, amended[33]), it was nonetheless the first measures of the national territory whose intention was to regulate land use in flood-prone areas.

---

32 See the Minister's report to the President of the Republic presenting the decree of October 30, 1935, JORF, p. 11688.

33 Particularly, the Seine, Aube, Marne, Meuse, Moselle, Loire, Allier, Rhône, Saône, Isère, Adour, Tech, Aude, Hérault, Var, Lot, Vidourle, etc.

The decree of August 29, 1955[34] specified the reasons for refusing building permits and concluded the turning point started by the law of 1935 regarding residential mobility constraints. This decree specified the application conditions of the urban planning code, particularly concerning *construction locations* (Paragraph 2), by considering (Article 2) that:

> "building on land exposed to natural hazards, such as flooding, erosion, subsidence, landslide, avalanches, may, if authorized, be subject to special conditions, considering that these lands are delimited by prefectorial order".

In the same way as the following texts on natural hazard prevention, this decree marked a clear evolution toward an anthropocentric and global approach in the area of hazards. The focal point now aimed at the control of human property exposure, while the decree law of 1935 had focused on avoiding obstacles to water flow, the SSP having favored the establishment of public use precautions and with that, the limitation of construction by default.

The law of July 13, 1982, regarding "the compensation for victims of natural disasters", established both Risk Exposure Plans (REPs) as well as insurance against natural disasters. The legal framework for the prevention of natural hazards was amended by the 1995[35] Environment Act. It established Risk Prevention Plans (RPP) which ultimately became the sole device for the prevention of natural risks, whatever their nature, by replacing all devices which had previously been in force (SSP) perimeter of the R 111-3 Article, from the Urban Planning Code, REP). RPPs had the same approach and purpose as REPs. These comprised (1) the delimitation of areas exposed to natural hazards, (2) the provision of necessary amenities and actions for ensuring people and property protection and (3) the control of urbanization and property development in areas exposed to risk. At the European scale, a few years later, directive 2007/60/EC passed on October 23, 2007 which regarded the assessment and management of flood risks, was based on a similar anthropocentric approach, particularly preventing the potential negative consequences of floods on human health, the environment, cultural heritage and economic activity, and sought to develop risk management plans in view of reducing these consequences. While Article 7.3 stated that flood risk management plans should embrace the totality of the managerial aspects (with a special emphasis on prevention, protection and

34 Decree no. 55-1164 of August 29, 1955, used as public administration regulation for applying Article 91 of the Urban and Habitation Code. JORF of September 3, 1955, pp. 8816, 8817.

35 Law no. 95-101 of February 2, 1995 on reinforcing the environment's protection, known as the Barnier law.

readiness), it made no reference to protection against flood expansion, nor any mention regarding the free flow of water.

The concern for the free mobility of water has not gone unnoticed in more recent legislation, though. Decree no. 93-351 passed on March 15, 1993 contemplated *specific provisions regarding the free flow of water and the conservation of flood plains* (Paragraph 2) and emphasized the role of the State to modify or remove obstacles for free flow. Article 4 of the 1993 decree regarding REPs required "justifying zoning criteria [...] both taking into account risk importance as well as land occupation or land use likely to create, aggravate or provoke new risks", thus acknowledging the singularity of floods in a very interesting way: the development that one desires to control can itself be an aggravating factor of vulnerability[36]. The 1995 the law also contained provisions on the free flow of water and flood plains. In this respect, the law modified article 16 from law no. 92-3 of January 3, 1992 regarding water, which specified that RPPs:

> "Define the necessary prohibitions and technical requirements to be respected in order to ensure the free flow of water as well as the conservation, restoration or extension of flood plains".

The law of July 30, 2003 paved the way for defining under prefectorial order public utility easement on pieces of land (1) adjacent to a watercourse or its derivation, or (2) part of a watershed or an estuarine zone, simply at the request of the State, the local authorities or other social groupings. These precautions could make it possible to create (1) temporary water retention areas in order to reduce flooding or runoffs in downstream areas and (2) "mobility areas of minor bed watercourses" upstream of urbanized areas in order to preserve or restore their essential hydrological and geomorphological features[37].

The residential and also the daily mobility of people has been addressed in various legislative texts in the form of forecasting and information measures – by giving information about risks in general, but also in times of crises, the legislator wishing to influence the behavior of people both in their residential and daily mobility. The law of 2003 was rich in prevention tools of this type, such as: the setting up of a master plan for flood forecasting (Article L. 564-2 of the Environment Code), organization of public municipal meetings (Article L. 125-2 of the same code), and maintenance of historical flood markings (Article L. 563-3 of the same code).

---

36 Article 2 of the law of 1935 defining SSPs was far more simplifying: "For valleys protected by dikes or heightened in any way, the plans do not take the existence of these works into consideration".

37 Article L. 211-12-II of the Environmental Code.

5.2.4. *Legal consideration for the multiscalar character of floods and the multiple uses of water (1964–2016)*

As it has been shown by the evolution of the legal settings evoked in previous sections, the question of floods interferes with different uses of water such as transport, energy production (1669, 1919[38]), agricultural development (1807, 1856) and pollution management (1964, 1992). It is obvious that these "accelerations" of the hydrological cycle entail certain implications, not only for the safety of people and property, but also for all human water-related activities. They concern broad scales that can range between agricultural plot or river reach, and the major river basins of a country.

In terms of legal provisions, it seems to us that taking into consideration the multiscale nature of problems related to the water cycle appeared in the law of December 16, 1964, which laid the first hydrologically structured bases for water management[39]. With the aim of fighting pollution, it chose to work according to different watersheds so as to economically and ecologically reconcile the different uses of water. Operating an unprecedented administrative organization in six major watersheds[40], this law opened the new perspective of institutional treatment of water-related problems on territorial entities adapted to the natural phenomenon. It acknowledged the structuring role of the hydrographic network, particularly concerning the interaction logic between upstream and downstream. Although avoided by the 1964 law, floods and risk-associated issues would benefit from this openness because they were deeply affected by this logic. All the issues raised over time around the questions of water flows, passive or active protection, and flood plain conservation, could only be understood by having a clear vision of the role of embedded basins, and could only be dealt with by mobilizing the most appropriate embedding levels.

38 Law on the use of hydraulic energy of October 16, 1919 (JORF of October 18, 1919, p. 11523–11528). This law responded to the hasty exploitation of the hydraulic resource put in place before and during the war because of a lack of access to the coal basins of the east. It basically fixed its concession scheme and its operating conditions. Article 10, paragraph 2 indicated that water regulation for company budgets established a balance between electricity production and other uses of water (agricultural, sanitary, tourist) as well as the protection against floods.

39 Law no. 64–1245 of December 16, 1964 in water regime and water distribution, as well as the combat against water pollution. In Paragraph 1, this law described the fight against water pollution and its regenerative mechanisms, and in Paragraph 2, the law referred to the water regime and the water distribution of waterstreams and subterranean water.

40 Six catchments were defined for the mainland territory: Artois-Picardie, Seine-Normandie, Rhin-Meuse, Loire-Bretagne, Adour-Garonne, Rhône-Méditerranée-Corse.

The law on water regulation of January 3, 1992, supplementing the 1964 law, introduced the principle of a balanced management of water as a resource, satisfying and reconciling different uses, activities or tasks and requirements, such as the conservation of free water flows, while granting protection against floods. This text finally meant a clear convergence between resource management, protection of aquatic environments and flood prevention. It adopted a hydrological logic of embeddedness and explicitly mentioned the notions of "subbasin grouping or subbasins corresponding to a hydrographic unit or an aquifer system", which was useful for establishing two levels of consultation between the State, the local elected representatives and users: (1) the Water Development and Management Master Plan (SDAGE[41]) which, developed as a result of the initiative by the Basin Coordinator Prefect, set as the fundamental orientations in resource management *for the whole basin or group of basins*, defined the qualitative and quantitative water objectives, as well as the adjustments to be made[42]; (2) the Water Development and Management Plan (SAGE[43]), which established a development and management plan for water resources and aquatic environments for a lower scale hydrographic unit[44] in view of preventing natural areas from turning into flood plains[45, 46].

In addition to the question of how to apprehend mobility, there is the question of the scale at which these modalities apply, as well as the way in which devices are managed.

41 *Schéma Directeur d'Aménagement et de Gestion des Eaux.*

42 Articles L 212-1 to L 212-2-3, R 212-6 to R 212-25 of the Environmental Code.

43 *Schémas d'Aménagement et de Gestion des Eaux.*

44 Articles L 212-3 to L 212-11, R 212-26 to R 212-48. The SAGE's perimeter was determined by the master plan; if this was lacking, it fell under the responsibility of the Prefect, after consultation or territorial collectivity proposal, and after consultation with the Basin's committee (Article R 212-26 of the same code).

45 Article L 212-5-1-4° of the Environmental Code.

46 The law on Water and Aquatic Environments (LEMA) of December 30, 2006 transposed Directive 2000/60/EC of October 23, 2000 to the water domain. It contained reforms to the legal framework established by the laws of December 16, 1964 and January 3, 1992. The LEMA mainly contained provisions concerning (1) the improvement of water governance management (including the creation of the National Office for Water and Aquatic Environments – Onema); (2) the conditions favoring a "good ecological condition" of water in 2015, imposed by the framework directive; (3) the simplification and strengthening of water police and public water and sanitation services; and (4) taking into consideration adaptation to climate change. Except for the obligation of a minimum flow imposed on the right of hydraulic works, it remained silent concerning the question of the common management of water as a resource and flood prevention.

5.3. The institutional scale and flood response

As we mentioned at the end of the previous section, the institutional consideration of the physical structure of hydrological systems responsible for floods is relatively recent. It appeared with the 1964 law, regulating the creation of water agencies and their usage fee system for regulating river pollution and water supply. The watershed framework naturally became the only relevant geographical unit for balancing the "polluter pays" principle[47]. We employ the term "naturally" because it is only by watershed and by subwatershed that the balance between different water uses and the amounts (and quality) of available water can be analyzed. This amount of available water is brought by rain to the surface of the basin and redistributed by infiltration and runoff into rivers, soils, vegetation and groundwater.

The question we are addressing here is that of the scale at which the institutional response to flood risk organizes the conflict between people mobility and water mobility. It is a question of comparing the hydrological embeddedness that governs water mobility, on the one hand, and the institutional levels capable of adapting people mobility, on the other. At what level a flood should be managed is an interesting question of subsidiarity, whose geographical dimension acts on the "mass" dimension[48]. Flooding is both a local problem of an individual – such as an owner exposed to a river reach – and a collective problem of population safety affected by a flooding water system.

In Table 5.2, we will show that the issue of public safety first emerged when it moved from the individual to the population. Later came the question of defining the appropriate level to deal with the problem, either at a municipal or at a national level. In both stages, the intersection of hydrologic embeddedness and "mass" levels that govern human relations as well as their territorial organization represented the crux of the flood problem. The solution proposed today is that of a negotiation between basin and subbasin responsibilities at an intercommunal level.

47 The "polluter pays" principle is subject to *Welfare Economics* and to A.C. Pigou, who considered pollution as a "negative externality" that should be limited by a tax. It is interesting that this principle entered the French legislation on water in 1964, before becoming, with the Barnier law of 1995, one of the four general principles of the French environmental law (prevention, precaution, participation – Article L110-1 of the Environmental Code).

48 "Mass" is one of the three dimensions employed by Grossettifor describing social phenomena [GRO 07]. This action scale describes a new level of interaction – from interpersonal to global.

Law's aim	*Administrative management level*	*Year*	*Hydrological object*
Protection against the action of water falls under the responsibility of resident owners and trade unions	Individual	1801	River reach and small basins
Owners must group themselves in Forced Trade Unions (FTU), then free unions	Trade union	1807	
		1865	
The municipal authority must grant the maintenance of the public order	Municipality	1789	
The municipal authority must "put an end […] to accidents and disastrous events"		1884	
The GEMAPI (*Gestion des Milieux Aquatiques et Prévention des Inondations* or Management of Water Environment and Flood Prevention) becomes an exclusive and mandatory intercommunality competence	Intercommunality and mixed trade unions	2015	Basins and subbasins
PZIs are managed by the Departmental Council	Department	1898	Valleys
Water as a resource is divided into subbasins (SAGE) – water management and distribution	Public basin agencies and establishments	1992	Basins and subbasins
The official managing actors are the public basin agencies and establishments (EPTB)		2003	
The water regime and water distribution are managed by Basin Agencies		1964	Great basins

Water as a resource is managed per basin (SDAGE) – aims, arrangements		1992	
Flood-prone Area Plans (PZIs) make it possible to manage flood encumbering objects		1858	
Building or planning in waterstream beds is regulated	National	1858 1935	Valleys
Water mobility is placed under the State's protection		1669	

Table 5.2. *Legislative evolution concerning floods in terms of administrative management. The main legal aims mentioned in section 5.1 have been resumed and classified depending on their administrative management level. The promulgation year of the law as well as the hydrologic level concerned are described in view of each object*

5.3.1. *From an individual, economic and sectorial concern to a public safety problem*

In rural France, "the elaboration of the first protection rules did not characterize a strategy, but the determination of a right adapted to the circumstances. The result was a sectorial protection of goods and people" [DRO 15]. Having more arable land or forest areas involved drying up large swamp areas and at the same time building protective structures, dikes, as a protective measure against the damage caused by floods, which was provided by the law of 1807. The public policy favoring the drying up of swamps was associated with the private promotion of the public maritime domain. The drying up of swamps that favored land appreciation, particularly those in the maritime public domain, was counterbalanced by the assumption of responsibility for the accompanying risks on the part of riparian owners [DRO 15], instituting the principle of protection against the action of water (against rivers or navigable or non-navigable torrents) at the expense of these owners, in the proportion to their interest in the works. At the beginning of the 19th Century, protecting oneself against floods thus became an individual responsibility and therefore, a sectorial action. Paradoxically, while flood protection became a public security problem at the end of the 19th Century, contemporary court decisions have confirmed the still valid application of such a principle, and affirm that neither the State nor municipalities are to be held accountable for flood protection of exposed property[49].

At the same time, in order to oblige riparian owners to build protective dikes and to maintain them, the laws of the 14th Floréal, Year XI and of September, 16th 1807 awarded public authorities with the power to compel owners to be grouped under forced trade union associations (*Associations Syndicales Forcées*) for the achievement of collective works (in particular, the construction of protective works against the action of water) as well as the maintenance of collective works. The law of June 21, 1865 introduced another possibility: the creation of free trade union associations (*Associations Syndicales Libres*), without administrative intervention, for the same purpose of carrying out collective work and the maintenance of works and facilities[50].

While the protection against the action of water was still a private issue, the safety of people was also a public mission and, in the wake of catastrophic events, flooding became a public concern and was the subject of provisions laying the foundations for a real national flood risk prevention policy.

The law of May 28, 1858 endowed the State with a leading role in the execution of works intended to protect cities from floods[51]. It was promulgated following a series of floods that hit Europe in the middle of the century and particularly after the floods of the Loire and the Rhône in May, 1856. In his explanatory memorandum,

49 For example, on June 10, 2010, the Administrative Court in Lyon, in the case of co-ownership of Aumard building center, no. 08LY01256, stated: "in the absence of legislative or regulatory binding provisions, municipalities do not have the obligation to ensure the protection of the neighboring properties of watercourses against water's natural action". On February 12, 2008, the Administrative Court in Bordeaux, MX, no. 05BX01268 declared that: "since there was no provision requiring the State to protect the properties bordering Ariège against the natural action of water, MX is unfounded to argue that the State might have been required to take all the necessary measures in order to prevent the deterioration of river banks".

50 The provisions regarding trade union associations now fall under ordinance no. 2004-632 of July 1, 2004.

51 The same is applicable to the laws of July 28, 1860 and June 6, 1864, known as Mountain Territory Restoration laws (MTR).

the President of the Parliament[52] began the presentation of this bill *on the defense of cities against floods* with:

> "Gentlemen, you must certainly have not forgotten the floods which, in 1856, devastated so many beautiful and rich regions of France" [CHA 61].

As the President observed:

> "The irregular, but inevitable, periodicity of these plagues which, after the exceptional floods of 1725, 1735, 1772 for the Garonne, 1840, 1841, 1856 for the Rhône, 1840 for the Saône, 1843 for the Durance, 1790, 1846 for the Loire, from 1740 for the Seine, sowed terror in these different valleys, particularly striking the people's mind".

It was therefore the repetition of large-scale events that led the authorities to transform flooding into a public concern, particularly with regard to the State's sovereign mission to ensure public security.

The decree law of 1935 was also an answer to a designated event, the floods of the Garonne, Aude and Herault of March, 1930. In their report to the President of the Republic, the President of the Council and several ministers started their *exposé* with these words:

> "Mr. President, the disastrous floods of the spring of 1930 have clearly highlighted the serious disadvantages resulting from [...] the congestion of the flood field"

With this decree law, flood prevention fully entered the sphere of public problems under the responsibility of the State. The State was endowed with the means to intervene on all the dimensions concerning the urban planning of submersible zones, but without turning its implementation into a doctrine, which was entrusted to the local administrative levels.

The following texts (Decree of 1955, laws passed in 1982, 1995 and 2003) would only reinforce this public concern about flood risk toward people and property. The shift from a private concern to a public issue entices us to explore the

52 In 1858, the legislative body, the Legislative Assembly of the Second Empire, was presided by the Duke of Morny.

reasons for this noticeable change with regard to the relevant administrative intervention level.

5.3.2. *Research for the pertinent "local" intervention scale*

Flood protection was a "local" concern before becoming a question of national interest. At the same time that the laws of the 14th Floréal, Year XI and September 16, 1807 were passed, thereby introducing the principle that the protection against the action of water would fall under the responsibility of riparian owners and the constitution of trade union associations for the purpose of executing protective works, the municipal law of December 14, 1789 conferred the municipal authority with the charge:

> "of allowing people to enjoy the benefits of an efficient police force, granting cleanliness, hygiene, safety, security and tranquility in the streets, public spaces and public buildings".

This law foreshadowed the provisions related to the attributions of the general police, which, after the law of April 5, 1884, regulated the municipal organization and required the municipal authority to grant the maintenance of public order, namely "to put an end to [...] accidents and calamitous plagues" (Article 97-6o. of the law[53] in question), and therefore, to maintain public safety[54]. Thus, the municipal

---

53 Provisions related to municipal police stated in the 1884 law were practically transposed word by word to Article L 131-2 of Municipalities Code, later becoming Article L 2212-2 of the General Local Authorities Code.

54 There is no contradiction between the principle of protection against the natural action of water at the expense of private owners, on the one hand, and the power of general police destined to ensuring the safety of citizens, on the other hand. Under this apparent contradiction appears the question of the intervention of public authorities in the field of public security and public interest. Thus, the magistrates of the administrative order consider that the Mayor is not compelled to intervene if the public safety is not threatened, and if the risk only concerns a property (Administrative Court of Lyon, May 3, 2012, Municipality of La Clusaz, no. 11LY00157, Administrative Court of Lyon, December 7, 1989, Mr. Marcel X and others, no. 89LY00397, 89LY00399, 89LY00400, 89LY00401). It may, however, compel a private owner to carry out works or to set up prevention devices on his plot (CE October 22, 2010, Mr. and Mrs. C., no. 316945, Administrative Court of Marseilles, February 9, 2004, Mr. René X., no. 00MA00511 – these two cases concern a risk of landslide). Nevertheless, in a situation of serious or imminent danger, the magistrates have adopted a constant jurisprudence, according to which they consider that the Mayor, in charge

police under the aegis of the city's Mayor was the first public authority in charge of ensuring security. Under this policing power, municipalities could become involved in the protection of citizens and undertake, as such, any tasks, works or installations proving to be of general interest or urgency[55].

The 1858 law transformed flooding into a matter of national concern, because the government issued an authorization for constructions or developments in watercourse beds, and made the decision to build dikes for the sake of the defense of the most prominent French cities. However, this law[56] reduced its practical implementation to a "local" concern, with plans for submersible areas at the scale of submersible valleys. In 1898, Book 2 of the Rural Code fixed the departmental level to manage PZIs and various objects, which could adversely disrupt flooding, implementing, to this purpose, the precautionary principle whose application was given to "the caution of the Regulations of the public administration [...] made after research and opinion collection from the Departmental Councils". The decree of 1955 delineating the perimeters of risk in Article R 111-3 of the Urban Planning Code, heirs of the PZI and the SSP, attributed to the State the load of delimiting the perimeters of risk and regulating the development of constructions in these perimeters via the authority of departmental prefects. From 1982 onwards, the REPs and then the RPPs took over the previous area delimitation, and took over national policies relayed at a departmental level.

The law of January 3, 1992 instructed each basin committee[57] to draw up at two different hydrological levels, a Water Development and Management Master Plan (SDAGE) and a Water Development and Management Plan (SAGE), which could

---

of ensuring public security, is obliged to carry out protection work, including on private property, for reasons of public safety (Administrative Court of Douai, February 22, 2007, Municipality of Gonfrevillel'Orcher, no. 06DA00494) and even when only a few people are threatened. In this case, works fall under the responsibility of the Municipality.

55 This obligation to ensure public safety has been the subject of numerous court decisions. See in particular EC, March 14, 1986, Commune of Val d'Isère, no. 96272. The general police power destined to ensure security in the territory of the municipality also implies, and *a fortiori*, an intervention on the part of the Mayor in case of emergency (in case of "serious or imminent danger"), in virtue of Article L 2212-4 of the General Local Authorities Code. The presence of a serious or imminent danger justifies the use of the mayor's policing powers, which may prescribe the execution of all kind of security measures required by the circumstances.

56 Like the decree law of 1935, which resumed the notion of Flooded Area Plans of the French great rivers and their main subsidiaries with the SSPs.

57 The basin Committees stemmed from the law of 1964 and its new organizational structure for large catchments (see section 5.1.4).

define the scope of local interventions and actions to be carried out in a more realistic way. A Local Water Commission (*Commission Locale de l'Eau* (CLE)), a consultative body for different water users established by the departmental prefect for a region where a SAGE was implemented, was responsible for leading the consultation process, defining its areas of work, for seeking financing means and for carrying out the operative enactment of the SAGE.

We observe that all these texts have struggled to identify the "local" level that is appropriate for the practical implementation of a national will to address the question of flooding. For this, successive levels are summoned, ranging from municipal levels to departmental levels, from basin valleys to subwatersheds, thus creating certain confusion between administrative levels of local authorities and management levels related to the material dimension of the phenomenon. Following the flood events of the late 1980s, this confusion on the definition of a "local" level has been continuously questioned through the reports of inquiry commissions, information missions, technical missions and others ones at the request of the Ministry of the Environment, the National Assembly, the Senate and the Court of Auditors.

5.3.3. *The formalization of the subsidiarity question*

The question of the division of competences and missions between the State, the local authorities and the riparian owners has clearly been posed as a question of subsidiarity, in our opinion, since 1997.

It was first posed in the study called the "Bourrelier" report, commissioned by the Ministry of the Environment [BOU 97], eager to go beyond *ad hoc* expertise reports to carry out a study on the effectiveness of the natural risk prevention policy. The comittee suggested refocusing the State's action in the determination, impetus and control of the natural risk prevention policy. With regard to local and regional jurisdictions, the authority considered that it was necessary to correct or compensate for the fragmentation of municipalities. The proposed subsidiarity levels are the grouping of municipalities under the aegis of intercommunalities with significant resources, intermediate collectivities (department, region) and institutions still to be created in order to coordinate separate territorial units, levels which, in the terms of the report, constituted "an expression of solidarity and territorial mesh[58]". This subsidiarity problem was doomed to systematic discussion in the various reports that

58 We would like to add the hydrologic mesh.

followed. In 1999, the "Daugé" report [DAU 99][59], more than a critical report on flood prevention, represented an inventory of fixtures and proposals for improving prevention and organization between the actors. Recommendations included administrative organization, by setting up a grid of major river basins, in partnership with the State and public territorial basin institutions, supported by all the administrative levels, particularly at the basin level (Coordinating Prefect and Directors of the Regional Environment Directorate[60], water agencies, public territorial basin institutions), by using all forms of cooperation between these bodies. The "Daugé" report recalled the role of the State, responsible for identifying risks, making them known, leading the risk prevention policy and monitoring the implementation of provisions. In 2000, the "Lefrou" report [LEF 00] pointed out the lack of collaboration between the State, the riparian owners and the local authorities, stressing the inadequacy of the legal system in the face of the situations to be managed. Proposals were intended to simplify this legal system and the distribution of competencies, implying a decentralization of these powers to the benefit of local authorities and their groupings, who were better suited to decide on the arrangements and to carry out maintenance and protection work – here, we re-encounter a typical search for subsidiarity, underpinned by budgetary concerns, with the State increasingly disengaging from the burden of local investment. Several reports [GAL 01, DEN 01, DOL 01, ANZ 10, ROU 11, NEC 12, COL 12, CAL 15] followed and all advocated for a decentralization of competencies and the relevance of entrusting flood management to the most appropriate local level, beyond traditional administrative limits, namely the department or the municipality. A carefully considered opinion of the State Corps [INS 02] showed that the whole of the events that took place in the 1990s brought about a clear awareness of the distinction between slow kinetic floods and flash (or "lightning") floods. From that moment onwards, State engineers explicitly expressed the multiscalar nature of the phenomenon as an element to be considered in prevention policies.

Since the beginning of the 2000s and in accordance with the recommendations contained in the reports mentioned, there has been a general movement of decentralization of State competences in the field of flood prevention. However, the State remains the competent authority for the elaboration of the RPPs; it takes part in the design of prevention plans and is the guardian of urbanization control in risk-prone areas. Melot and Bransiecq [MEL 16] affirm that in urban planning, the State is relegated to the position of guardian of the legality and arbitrator between the

59 Concerning the reflections involved in the framework of the "Bourrelier" report.

60 DIREN (*Direction Régionale de l'Environnement*), later transformed into the DREAL (*Direction de l'Aménagement et du Logement*).

different planning "layers" (communal, supracommunal) – a similar observation can be made in the field of flood risk prevention.

Local and regional authorities, municipalities and intermunicipal authorities are in charge of preventive actions and works. Local authorities may undertake any works, actions, protection works or installations having a general interest or sense of urgency within the framework of the Water Development and Management Plan (SAGE)[61]. In order to finance work expenses, they can engage the people who have made the works necessary or who have a special interest in them, particularly flood-exposed riparians[62]. They can constitute a fluvial public domain by transfer of the domain of the State[63]. The territorial public basin agencies and establishments (EPTB) have seen their role strengthened and are defined as[64]:

> "A group of local collectivities gathered at the level of a basin or group of sub-watersheds, destined to facilitate flood prevention and the defense against the sea, the balanced and sustainable management of the water resource, as well as to preserve, manage and restore the biodiversity of aquatic ecosystems and wetlands and to contribute,

61 Article L 211-7 of the Environmental Code, amended by laws no. 2003-699 of July 30, 2003 and no. 2006-1772 of December 30, 2006, the ordinance of May 6, 2010 – the list of works, action works and facilities includes control of rainwater and runoff, defense against floods and against the sea and hydraulic facilities contributing to civil security.

62 Articles L 151-36 to L 151-40 of the Rural Code and Sea Fishery, which also make it possible to avoid public inquiries for those works that are necessary for facing situations of imminent danger, provided that these works do not entail any kind of expropriation nor does the contracting authority intend to request financial participation from interested parties. Provided that they do not involve any expropriation nor any request for financial participation of the interested persons, a public inquiry may happen to not be required when the works that concern a watercourse covered by a SAGE are directly related to a flood, which has been declared as a natural disaster. These tasks must be completed within three years of the recognition of the natural disaster and must aim to restore the watercourse following its natural features.

63 Law no. 2003-699 of July 30, 2003, instituting Articles L 3113-1 to L 3113-4 of the General Code on the Property of Public Persons that enables territorial collectivities or their groupings to establish a framework of local intervention in terms of the management, development and control of watercourses. Articles R 3113-1 to R 3113-7 specify the methods for a coherent management of the transferred domain and mention a list of watercourses that cannot be subjected to such a transfer.

64 Article 46 of the law of July 30, 2003 that endows EPTBs with the status of official actor being involved in the balanced management of water and flood prevention at the scale of a watershed or a subbasin.

where appropriate, to the development and monitoring of the SAGE[65]".

It is the basin Coordinating Prefect who creates the EPTBs by determining the hydrographic entities that justify the creation or modification of such a perimeter.

Once this range of possibilities has been established in order to distribute the decentralized competences of the State to the "local" level, the question of the choice of the proper level of responsibility remains between three or four open "local" levels thus created. Once again, it was the coincidence of two major disasters in 2010, Cyclone Xynthia, which flooded several municipalities of the Atlantic façade and the stormy system that struck several municipalities in the Var department, as well as the legal reports and actions that followed, which led the State to adapt its legislation.

Reports that followed after Cyclone Xynthia and the Var floods of 2010 confirmed the lack of coherence of preventive actions, and, in the case of Cyclone Xynthia, the lack of governance concerning the management of protective works. In order to avoid future fragmentation of responsibilities for river management and flood control, and to favor the emergence of a coherent flood risk prevention policy, decentralization laws no. 2014-58 of January 27, 2014 (Law on Modernization of Territorial Public Action and Affirmation of Metropolises, known as MAPTAM) and no. 2015-991 of August 7, 2015 (on the new territorial organization of the Republic) transferred jurisdiction over the management of aquatic environments and flood prevention (known as the GEMAPI competence) to intercommunalities (metropolitan areas, urban communities, agglomeration communities, municipality communities) from January 1, 2018[66]. The GEMAPI competence became an exclusive and mandatory competence of intercommunalities, specified by the L 211-7 Article of the Environmental Code. These are competent for becoming involved in the development of watersheds, the maintenance and development of rivers, canals, lakes and bodies of water and the defense against floods and against the sea[67]. From

65 Article L 213-12 of the Environmental Code, stemming from law 2016-1087 of August 8, 2016.

66 It is possible for territorial collectivities to anticipate this competence transfer without having to wait until January 1, 2018.

67 Amendments to the Environmental Code facilitate dike management or prevention works by local authorities. Once their usefulness is confirmed, the old public-law dikes, whatever their initial owner, are made available to the local authorities in charge of flood prevention (Article L 566-12-1-I). The principle of availability is extended to works and infrastructure belonging to legal entities governed by public law, whose characteristics and location make

now on, the local authority responsible for water management and flood prevention is clearly identifiable.

The gamble made by these laws implied that intercommunality was the closest competent entity at the hydrological scale for managing floods. Indeed, water management on a scale exceeding the municipality's perimeter made it possible to improve the cohesion of policies and implemented actions by embracing a set of so-called "suburban" small-sized basins (2–20 km^2), which could be susceptible to flash floods. In addition, the law of August 7, 2015 established cooperation tools between intercommunalities, which facilitated the management of flood prevention for larger basins (20–200 km^2), which could also be affected by flash floods. For example, EPTBs were reorganized into mixed syndicates, particularly in view of coordinating actions within the framework of the GEMAPI[68]. Finally, in coastal river watersheds subject to recurring floods, or in the subbasins of a major river, a Public Water Management and Development Establishment (*Établissement Public d'Aménagement et de Gestion des Eaux* or EPAGE), that is to say, a mixed syndicate specializing in GEMAPI, was set up so as to bring intercommunalities together on an intermediate hydrographic scale (200–2,000 km^2)[69].

These recent laws therefore represent the legal consecration of water and flood management by interlocking administrative units designed for covering several basin scales without breaking up the competence. More generally, we can refer to the emergence of a coherent "functional" territory, a local action "territoriality" or even new "institutional territories" [LOR 11]. Through the control of urban development, whose competence falls within territorial collectivities[70], these different levels of

them a useful contribution regarding prevention (Article L 566-12-1-II). In addition, the competent local authority for flood prevention may establish precautions on the basis of former private dikes so that these structures are not destroyed and so that the collectivity can reuse them as part of the flood containment system it has chosen (Article L 566-12-2).

68 This coordination presupposes a competence transfer between intercommunalities or a delegation of these competences in cases where local authorities responsible for GEMAPI enforcement are not members of the EPTB (Article L 213-12 of the Environmental Code).

69 Article L 213-12 of the Environmental Code.

70 After the decentralization laws of January 7, 1983 and July 18, 1985, local authorities, namely municipalities, became competent for drawing up planning documents. Equally, for both urbanism and for risk, we come across the same search for relevant scale: at the intercommunal level, after the law of December 13, 2000, Master Plans became Diagrams of Territorial Coherence (SCOT); at the communal level, Land Use Plans (*Plans d'Occupation des Sols* or POS), which the same law later reformed into Local Urban Plans (*Plans Locaux d'Urbanisme* or (PLU)). The latest legal developments have turned the intercommunal level, the preferred level for defining urban planning projects. Thus, law no. 2014-366 of March 24, 2014 provided for a competence transfer in urban planning before March 27, 2017 to municipality communities and to existing agglomeration communities at the publication date

coherent intervention should make it possible to better manage residential as well as people's daily mobility. Beyond these subsidiarity considerations related to the flooding phenomena, this new stage of decentralization also revealed the State's willingness to disengage itself from a certain number of public expenditures, works or investments which now had an exclusively local, but no longer national interest.

5.4. In conclusion, a thread of founding ideas

Clearly focused on the legal institutional response, its development over time and its multidimensional character, this chapter also retraced the normative and cognitive institutional evolution, which accompanied this legal response. Behind the various legislative and regulatory phases, we have witnessed the development of specific knowledge regarding the phenomenon of flash floods, together with the problems it poses not only in terms of temporality, but also spatiality (cognitive point of view), as well as evolution in the way of envisioning the phenomenon as a danger to be fought or accepted (normative point of view).

Here, we reshape some elements regarding parallel institutional evolutions using the timeline in Figure 5.1. This timeline shows the main evolutionary phases of the

of this law or created or resulting from a merger of communities between the date of publication of the law and March 26, 2017. Any new municipality community or agglomeration community created after March 26, 2017 was compulsorily attributed this competence. Henceforth, land use planning documents would have to be elaborated at an intercommunal scale (Intercommunal PLU (IPLU)) which increased the size of the administrative units and favored a better coherence for intervention scales when dealing with floods and urbanization control in flood-prone areas.

In practical terms, the transition to these new scales has not been smooth. Law no. 2003-590 of July 2, 2003 (Town Planning and Habitat) enabled municipalities with a SOP in force before the promulgation date of the law of December 13, 2000, to keep it. Law no. 2014-366 of March 24, 2014, known as the ALUR law, made Land Use Plans invalid and would not turn them into PLUs by December 31, 2015, unless a Land Use Plan revision procedure was initiated before December 31, 2015. In this case, municipalities had until March 27, 2017 to make the change. After the reform introduced by the law of December 13, 2000, municipalities with weak urban development also had the possibility of drawing up a communal map, in view of determining town planning orientations. All these documents, after the law of July 22, 1987 on the organization of civil security, the protection of forests against fire and the prevention of major risks, were compelled to take risks into account in order to establish the constructability of specific areas.

previously described legal institutional response and groups them around four questions that we deem founding[71].

Flooding and water and river uses

This first question has been permanently present during the period considered, where it produces regular socioeconomic evolution markers (for the period designated as Caption (1) in Figure 5.1). The drying-out of swamps, land drainage and urbanization (indirect uses of water or at least of waterbeds and watercourses), transport, energy production and pollution (direct, quantitative and qualitative uses – in competition with food, hygiene and ecological concerns – over time) all represented different challenges to the freedom of water, particularly to the river but in no way limited to that, of naturally coming and going. These disputes were questioned in a visible way at the time of floods or low flows, that is to say, in extraordinary times of system functionning.

Flooding, individual protection and cooperation

Compared to other hazards such as earthquakes, the difficulties posed by flooding are due to the fact that protection and special arrangements modify the hazard. In particular, buildings are not only vulnerable themselves, but become vulnerable to each another – this is the conservative aspect of the phenomenon in that, due to the conservation of the water mass, property protection for some may result in hazards to other's property. The flood has to be somehow "shared", thus becoming a collective problem whose solution involves a certain form of cooperation.

These vulnerability transfer mechanisms from one owner to another, from one riparian community to another, from a city to suburban or rural areas along the same river, had to be identified in order to endow floods with the status of a public safety problem (period designated by Caption (2) in Figure 5.1).

Flooding and mobility

This question is at the heart of this book. Like the use of water, it has undoubtedly been present throughout the studied period, since flooding represents, by definition, the unexpected meeting in space between water and human activity. The first part of this chapter presented the studied period as a shift between the constraint of watercourse mobility and people's constraints. Nevertheless, we consider that this change did not really start until the middle of the 19th Century,

71 These questions concern sociohydrology, which seeks to understand the dynamics of the dyad water society [SIV 12]. The assumed hypothesis is that of including the human race in the study of the water cycle (for this, see Chapter 8).

when the problem of flooding was recognized as such, and that Flood Zone Plans appeared. These plans tried to define a boundary between water mobility and people mobility (period designated by Caption (3) in Figure 5.1).

Obviously, the first half of this period (at least until the beginning of the 20th Century) was marked by the notion of progress and scientific rationalism, all of which inferred that a better knowledge of the hazard would make it possible to control it either in an active or a passive way, and then to get organized accordingly. It would be necessary to wait until the pairing between "flood zone" and "building zone" is made (in the middle of the 20th Century) so as to bear witness to the turning point regarding the implementation of constraints to human property (housing, activities), this implantation being key to residential mobility.

Territory and hydrographic network – an original subsidiarity problem

The question of finding the appropriate level of cooperation in order to manage flooding is of the utmost importance for the studied period. The question was elicited as early as it became necessary to manage dikes, but it was only from the moment after residential mobility in floodable zones was controlled that deeper research was triggered (period designated by Caption (4) in Figure 5.1). This research would eventually overcome the lack of correspondence between the administrative division of the territory and the structure of the hydrographic network. It was not until 1964, with the creation of the Water Agencies that this question was formalized in terms of the singular subsidiarity problem. The law of July 30, 2003 officially recognizes EPTBs and the MAPTAM law of 2014 (which attributes the management of water and aquatic environment, as well as flood prevention to intercommunality authorities) reinforces the decentralization movement in this field, together with subsidiarity.

Today, flood prevention has become a fully integrated issue to the greater question of water management, itself a crucial element for achieving full sustainable management. Although Article 33 of the 1807 law is still in force (making the owners protect themselves from the action of water), flooding is still, above all, a matter of public security, having very recently led to clarifying the relevant administrative level that should intervene for preventing this risk and for avoiding future developments, which may modify water flows, and thus aggravate flooding effects. Also, other actors are involved in the management of the link between water mobility and the mobility of people. Other chapters of this book complement this one, and make it possible to make a global analysis of the mobility problem, in particular through the question of urbanization control, or still through the mechanisms against flood exposure.

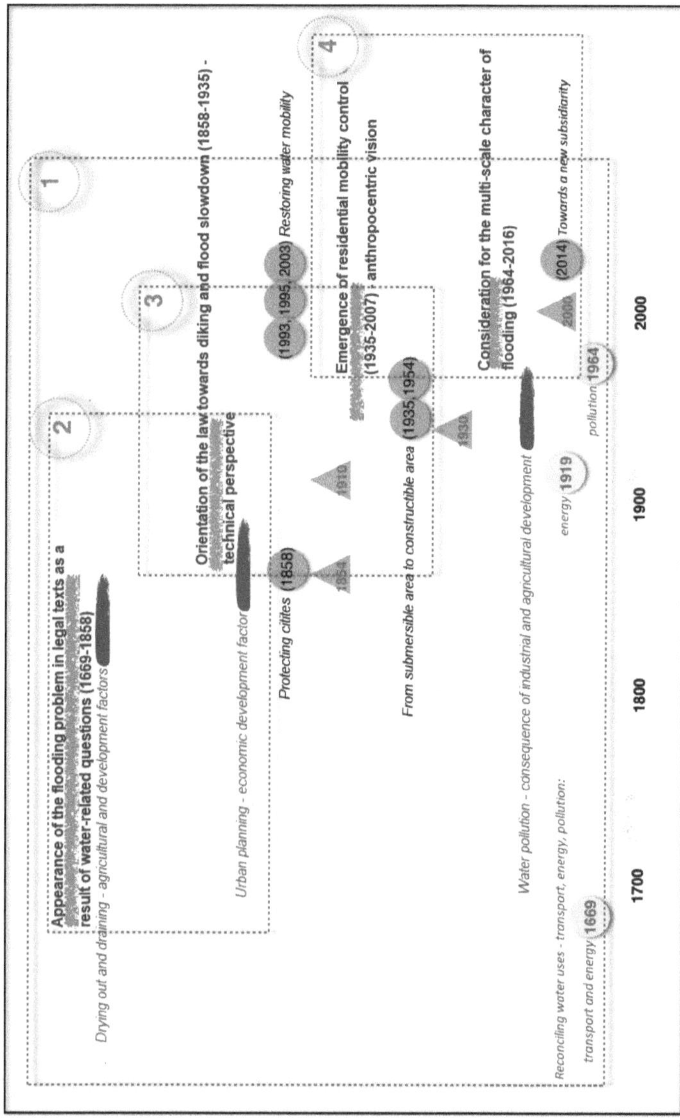

Figure 5.1. *Diagram of the French institutional evolution related to flooding in the last three centuries. This timeline stresses the great legal institutional evolution phases described in section 5.2 (gray shadows), in relation to catastrophic floods having a national impact (blue triangles), legal promulgation (flood-related – blue circles and related to energy production, transport... – yellow circles), and finally in relation to society's evolution with regard to water (drainage, urbanization, pollution... – green shadows). Four boxes (gray dots) refer to the founding questions we considered in section 5.4: (1) water uses; (2) solidarity; (3) mobility; (4) subsidiarity. For a color version of this figure, see www.iste.co.uk/lutoff/hydrometeo1.zip*

5.5. Bibliography

[ANZ 10] ANZIANI A., Xynthia : les leçons d'une catastrophe, Report no. 554, Sénat, June 10, 2010.

[AST 75] ASTRIE T., *Les drames de l'inondation à Toulouse*, Hachette BNF, Paris, 1875.

[BEU 08] BEUCHER S., REGHEZZA-ZITT M., "Gérer le risque dans une métropole : le système français face à l'inondation dans l'agglomération parisienne", *Environnement Urbain/Urban Environment*, vol. 2, pp. 1–10, 2008.

[BOU 97] BOURRELIER P.-H., DENEUFBOURG G., GRASSIN J. *et al.*, *La prévention des risques naturels. Rapport de l'instance d'évaluation*, La Dcumentation française, Paris, 1997.

[CAL 15] CALVET F., MANABLE C., Xynthia, 5 ans après : pour une véritable culture du risque dans les territoires, Report no. 536, Sénat, June 18, 2015.

[CHA 61] CHAMPION M., *Les inondations en France depuis le VIe siècle à nos jours*, Dunod, Paris, 1861.

[COL 12] COLLOMBAT P.-Y., Se donner les moyens de ses ambitions: les leçons des inondations du Var et du sud-est de la France, Report no. 775, Sénat, September 24, 2012.

[COU 00] COUR DES COMPTES, Rapport public annuel sur les comptes de 1999, Report, 2000.

[COU 12] COUR DES COMPTES, Les enseignements des inondations de 2010 sur le littoral atlantique (Xynthia) et dans le Var, Report, 2012.

[DAU 99] DAUGE Y., Les politiques publiques de prévention des inondations, Report, La Documentation française, 1999.

[DEN 01] DENEUX M., MARTIN P., "Les inondations de la Somme, établir les causes et les responsabilités de ces crues, évaluer les coûts et prévenir les risques d'inondations", vol. 2, Report 34, Sénat, October 23, 2001.

[DOL 01] DOLIGE E., Les inondations de la Somme, du Cher et de la Vilaine, rappel des inondations précédentes, Assemblée nationale, Report, 2001.

[DRO 15] DROBENKO B., "Risques littoraux : entre stratégie et droit", *Vertigo*, Special edition, 2015.

[FON 08] FONTAINE P., "Qu'est-ce que la science? De la philosophie à la science : les origines de la rationalité moderne", *Recherche en soins infirmiers*, vol. 92, pp. 6–19, 2008.

[GAL 01] GALLEY R., FLEURY J., Les causes des inondations répétitives ou exceptionnelles et les conséquences des intempéries afin d'établir les responsabilités, d'évaluer les coûts ainsi que la pertinence des outils de prévention, d'alerte et d'indemnisation, Assemblée nationale, Report, 2001.

[GAS 54] GASPARD B., "Les crues de la Seine à Paris", *La Houille Blanche*, Special edition, 1954.

[GRO 11] GROSSETTI M., "L'espace à trois dimensions des phénomènes sociaux", *SociologieS*, 2011.

[GUE 14] GUERRIN J., Une inondation négociée ? Politisation d'un risque naturel sur le Rhône, PhD Thesis, Montpellier I, 2014.

[HOF 99] HOFFMAN A.J., "Institutional evolution and change: Environmentalism and the US chemical industry", *Academy of Management Journal*, vol. 42, pp. 351–371,1999.

[INS 02] INSPECTION GÉNÉRALE DE L'ADMINISTRATION, DU CONSEIL GÉNÉRAL DES PONTS ET CHAUSSÉES, DU CONSEIL GÉNÉRAL DU GÉNIE RURAL, DES EAUX ET DES FORÊTS, INSPECTION GÉNÉRALE DE L'ENVIRONNEMENT, Les retours d'expérience des inondations catastrophiques et les inspections des services déconcentrés en charge des risques naturels, Report, June 3, 2002.

[LEF 00] LEFROU C., MARTIN X., LABARTHE J.-P. *et al.*, Les crues des 12, 13 et 14 novembre 1999 dans les départements de l'Aude, de l'Hérault, des Pyrénées-Orientales et du Tarn, Report, La Documentation française, October 2000.

[LEM 12] LEMARCHAND G., "Communication au sujet d'une dépêche de M. le Ministre des travaux publics relative aux travaux de défense contre les inondations", vol. 3, Conseil Municipal de Paris, 1912.

[LOJ 82] LOJKINE J., "Crise et renouveau de la sociologie du travail : À propos du paradigme techniciste", *Sociologie du Travail*, vol. 24, pp. 192–206, 1982.

[LOR 11] LORANT-PLANTIER E., PECH P., "La gestion du risque inondation en France vecteur de territorialité", *Annales de géographie*, vol. 678, pp. 193–203, 2011.

[MAT 94] MATHOT P., MARIANI T., Les causes des inondations et les moyens d'y remédier, Assemblée nationale, Report, 1994.

[MEL 16] MELOT R., BRANSIECQ M., "Règles d'urbanisme et choix politique : les observations de l'État sur les projets locaux", *Revue d'économie régionale et urbaine*, vol. 4, pp. 767–798, 2016.

[MON 58] MONESTIER-SAVIGNAT A., *Étude sur les phénomènes, l'aménagement et la législation des eaux au point de vue des inondations avec application au bassin de l'Allier*, Victor Dalmont, Paris, 1858.

[PON 89] PONTON A., DENIEL J., ESTIENNE J. *et al.*, Mission technique chargée de tirer les enseignements de la catastrophe de Nîmes du 3 octobre 1988, Report, Délégation au risques majeurs, Neuilly-sur-Seine, 1989.

[REG 06] REGHEZZA M., Réflexions autour de la vulnérabilité métropolitaine : la métropole parisienne face au risque de crue centennale, Thesis, Paris X-Nanterre, 2006.

[ROD 10] RODE S., "De l'aménagement au ménagement des cours d'eau : le bassin de la Loire, miroir de l'évolution des rapports entre aménagement fluvial et environnement", *Cybergeo: European Journal of Geography*, available at: http://cybergeo.revues.org/23253, 2010.

[ROU 11] ROUZEAU M., MARTIN X., PAUC J.-C., Retour d'expérience des inondations survenues dans le département du Var les 15 et 16 juin 2010, Report, La Documentation française, 2011.

[SIR 98] SIREY J.-B., *Recueil général des lois et des arrêts : en matière civile, criminelle, commerciale et de droit public*, 1er Cahier, 1898.

[SIV 12] SIVAPALAN M., SAVENIJE H.H.G., BLÖSCHL G., "Socio-hydrology: A new science of people and water", *Hydrological Processes*, vol. 26, pp. 1270–1276, 2012.

[TRO 47] TROTÉ J., *Traité des eaux non domaniales*, Imprimerie Nationale, vol. II, Paris, available at: http://www.ghzh.fr/, 1947.

[VAL 59] VALLES M., Mémoire sur les inondations, Annales des ponts et chaussées. Mémoires et documents relatifs à l'art des constructions et au service de l'ingénieur, Text, 1859.

6

The Difficulties in Local Implementation of Risk Prevention Policies

6.1. Introduction

Flooding is the leading natural risk in metropolitan France: almost half of all communes are at risk[1]. In 2011, a little over 17 million people lived in flood-prone areas (25% of the population) and nine million worked in flood-prone areas (one out of three jobs). Recent decades have seen rapid urbanization, including in areas known for their flood risk. When these areas combine an accentuated topography with intense and violent rainfall, very sudden and devastating flooding can occur. Faced with these catastrophic situations, prevention work has been carried out to protect against such risks.

Legal regulations have also been issued with the main objective of controlling urbanization in risk areas and reducing the vulnerability of goods and people. However, although the potential for public authority intervention has been there since 1935 (with the establishment of the "Plans de Surface Submersible" – PSS – Submersible Surface Plans), resorting to different mechanisms to control urbanization in flood areas remained rare until the establishment of the *Plans de Prévention de Risques Naturels Prévisibles* (PPRN – predictable natural risk prevention plan) in 1995, which substituted the previous mechanisms[2].

Chapter written by Joël BOULIER, Edwige DUBOS-PAILLARD, Antonin PAVARD and Valérie SANSEVERINO-GODFRIN.

1 According to the *Commissariat général au développement durable* (French secretariat general for sustainable development), 2014.

2 See Chapter 5.

This weak implementation of regulations may seem surprising since existing regulations were included in urban planning documents. Since the French law of July 22, 1987, all urban planning documents must take natural risks into consideration. Mayors can also intervene to limit the exposure of goods and people. The *Schéma de Cohérence Territoriale* (SCOT – territory coherence scheme), *Plans Locaux d'Urbanisme* (PLU – local urbanism plans) and even communal maps, respectively, set the urbanization development directions and objectives and the rules of ground usage, and determine the constructability of areas in terms of natural risks. Due to these specific natural risk prevention dispositions as well as those linked with urban planning rights, urbanization areas have therefore been determined, in principle, in terms of flood-prone areas.

The establishment of *Plans de Prévention des Risques Naturels Prévisibles* is often presented as a turning point in terms of taking risk into account. However, the effects of PPRI implementation were not the strict prevention of urbanization in flood-prone areas. Certainly, the different risk zones of the *Plans de Prévention du Risque Inondation* (PPRI – flood risk prevention plan) are based on the assessment of hazards (definition of baseline tides, hydraulic modeling) and the stakes, but they also take into account the local economic and social specificities. They are the fruit of negotiations between the State and local actors [MOU 11].

In fact, until the mid-1990s, it has been noted that the different procedures that enable the limitation of construction in risk areas were rarely used by the administration. Although the awareness of risks and new legal provisions allowed the limitation of construction in flood-prone areas, increasing urbanization continued largely, despite everything, in areas of flood risk.

Negotiated locally, the pursuit of urbanization in flood-prone areas was not forced. It remains to be seen how the major floods that marked the last 25 years have influenced both the rate of urbanization and the implementation of regulatory measures.

Our work consisted of putting the coexistence of regulations and urbanization into context at a territorial level. For this, we chose to observe how this was expressed in the Gard department, France. The choice of the Gard is based on the exemplary aspect of this department:

– this territory is specific in the sense that it recurrently and over long time periods undergoes "flash flood"-type flooding which generates considerable damage;

– there are many natural disaster bylaws (CATNAT) there: all of the communes of Gard have had at least one CATNAT classification and more than 40% of them have had more than four between 1982 and 2016;

– despite a highly significant and mediatized flood context, 35% of the population is exposed in flood-prone areas, and urbanization in flooding areas has continued to progress.

Through this example, we initially researched whether significant floods influenced the rate and location of recent urbanization, and whether they were able to lead the evolution of legislation in terms of urbanization in flood areas. Next, we tried to understand how urbanization in flood areas took into account, in a differentiated manner, the legal constraints from one territory to another.

6.2. Evolution of legal provision: urbanization versus flooding?

The evolution of legal measures in terms of risk protection are the reflection of a duality: on the one hand, the evolution of demography or of economic needs have led to the pursuit of urbanization, and on the other hand, another side of social demand has pushed the legislator to suggest legal measures that are better adapted to the prevention and protection needs of goods and people faced with flooding. This is why, on one hand, the evolution of legal measures is marked by a certain coexistence between flooding events and urbanization; on the other hand, this evolution is also partially triggered by major traumatic events.

6.2.1. *Versatile coexistence between urbanization, flooding and legal provisions*

6.2.1.1. *Legal provisions born out of a context of continued urbanization*

Deadly floods in France have particularly marked the second half of the 20th Century. Mediterranean regions are particularly affected, notably due to the population increase and its concentration in and around cities that are largely located in valleys or on the coast. As shown in Figure 6.1, we note an acceleration of dramatic events in the 1980s. We also see that the more greatly impacted territories are heavily urbanized and dense. These territories are particularly present in the Mediterranean arc.

Despite historic knowledge of these floods and the damage generated, urbanization was not completely interrupted. According to Renaud Vié Le Sage, the previous director of the major risk commission, "in France, 80% of building permits in flood areas were issued during the last 40 years [between 1950 and 1990]"[3]). In mid-Loire (between Nevers and Angers), the exposed population went from 30,000

---

3. From the magazine *50 millions de consommateurs* (50 million consumers), 1993, quoted in the National Assembly, 1994.

people in the mid-19th Century, to 120,000 around the year 1950, and to 300,000 in the early 2000s [DEG 02]. In Gard, 278,000 people lived in flood areas in 2006 compared to 234,000 in 1990.

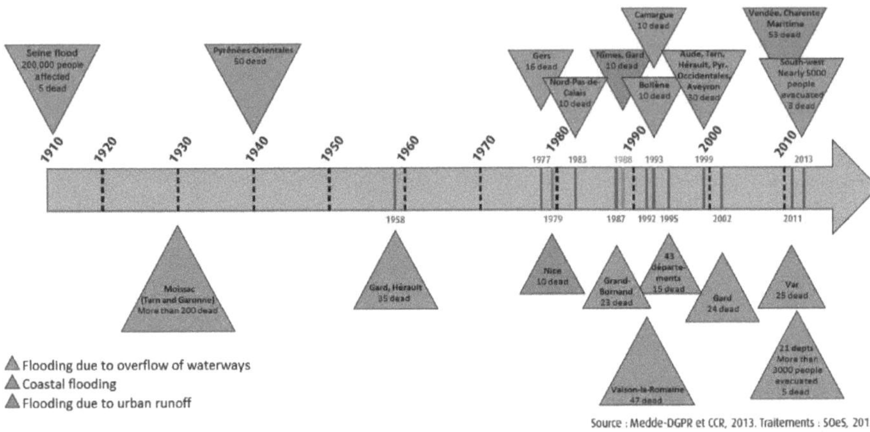

Figure 6.1. *The most deadly floods that have occurred in France since the early 20th Century (source: [FRA 14]). For a color version of this figure, see www.iste.co.uk/lutoff/hydrometeo1.zip*

Since the 1950s, legal provisions with regard to flood risk have been established by the legislator in this context of continued urbanization. However, the absence of dramatic events in the 1960s and 1970s, due in part to the completion of many developments on the main waterways (containment, construction of reservoir basins), led the majority of representatives to respond primarily to the strong social demand of creating accommodation and premises for activities, at the expense of the necessary protection of individuals facing the risk of flooding.

Is the rate of urbanization marked by the risk of flooding?

Gard is a department emblematic of the difficult establishment of legal measures. It is one of the French departments that are most exposed to the risk of flooding. Almost all communes in Gard are exposed. From the mid-13th Century to today, more than 500 floods have been recorded (*Conseil Départemental du Gard*, 2016). Amongst the most recent, those in 1958 (Vidourle, Gardon, Cèze), 1988 (Nîmes), 2002 (Cévennes and Rhône) and more recently, 2014 (Vidourle, Gardon, Cèze, Vistre-Rhony), led to considerable material damage and often caused human casualties (Table 6.1).

Date of flood	Number of communes affected	Number of deaths	Estimated material damage
1958	64	36	Undisclosed
1988	76	11	€610 M
2002	299	25	€830 M
2003	37	1	€300M (in Gard)
2005	86	0	€26.65M
September–October 2014	158	5	Undisclosed

Table 6.1. *Major events that have affected Gard since 1958 (source: NOE Gard 2016)*

The majority of floods are due to intense and sudden rains that often fall in autumn, on the small watersheds with steep slopes, leading to torrential flooding. Floods that are devastating due to their speed and force are the most frequent in this department. Thus, the 2002 flood (25 dead, 299 communes affected, 860 million euros worth of damage, 7,200 homes and 3,000 enterprises affected, according to the prefecture of Gard) constituted for many communes, including Gardons and Vidourle, a landmark event due to its exceptional pluviometry and hydrological characteristics. The East of the department is more exposed to the slow floods of the Rhône, similar in their characteristics to those of the Parisian basin, for example. Less brutal, their impact can nevertheless be considerable (as was the case, for example, during the events of 1993, 1994, 2002, 2003, etc.).

In Gard, more than 121,000 ha of land is flood prone, equivalent to 21% of the department's total surface area. The sectors that are most affected by floods are, logically, often the most populated, positioned mainly in valleys, as seen with the floods in Cèze (Bagnols-sur-Cèze), Gardon (Alès), Vidourle (Sommières) or in Vistre and Rhône (Figures 6.2 and 6.3).

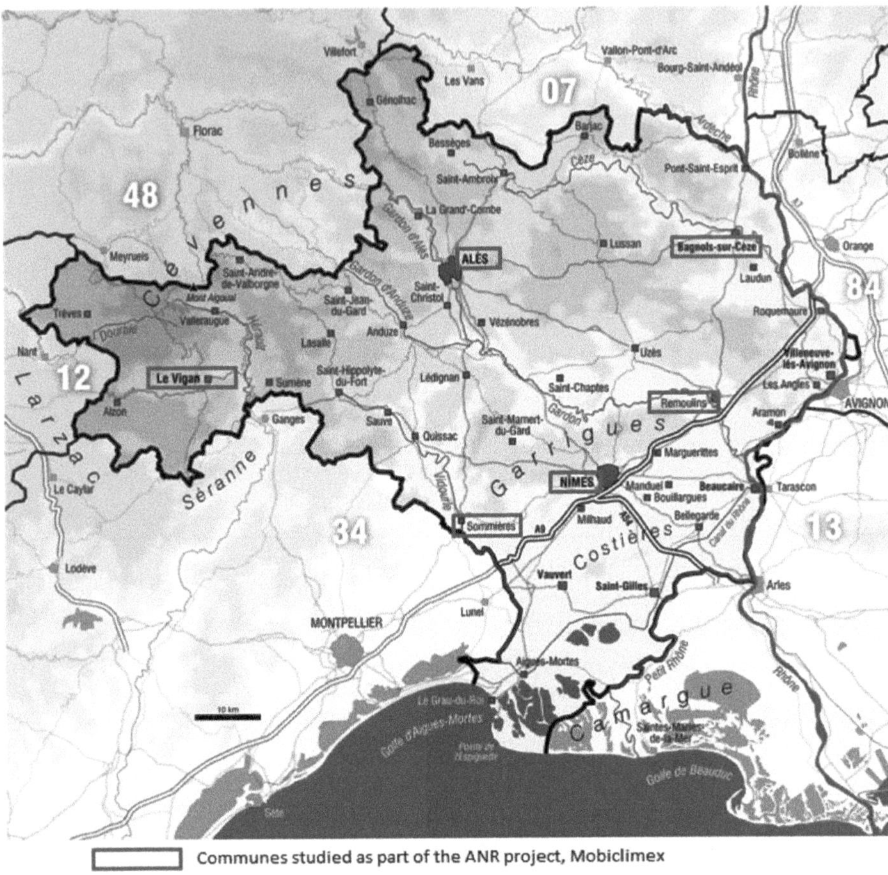

Figure 6.2. *Gard, between Vidourle, Cèze, Vistre and Gardon(s). For a color version of this figure, see www.iste.co.uk/lutoff/hydrometeo1.zip*

The largest communes in Gard are mostly situated in flood plains (Figure 6.3). A large population is therefore regularly exposed to recurrent flooding. Although Gard's history is marked by numerous floods[4], a hydrological break was nevertheless noted between 1958 and 1988 [VIN 10]. During this same period, the department experienced an important demographic growth combined with a concentration of the population in the main agglomerations. Thus, between 1968 and 2011, the population multiplied by 1.5 and the number of homes more than doubled (source: RGP-INSEE).

4 See the inventory carried out by the Gard departmental services: http://www.noe.gard.fr/index.php/observatoire-du-risque-inondation/indicateurs?id=56.

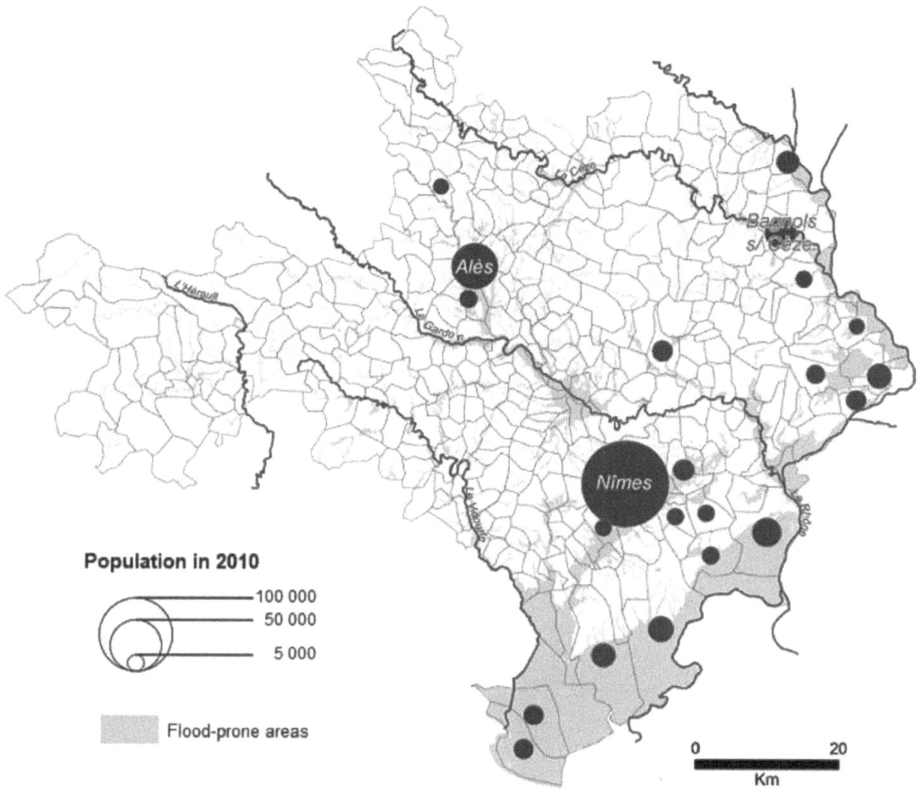

Sources : Insee, RP 2010; NOE - conseil départemental du Gard, zones inondables *Antonin Pavard, 2016*

Figure 6.3. *Reference flood-prone areas (CD30, 2013) and communes of more than 5,000 inhabitants in Gard. For a color version of this figure, see www.iste.co.uk/lutoff/hydrometeo1.zip*

However, due to a lack of standardized data, it remains difficult to quantify the evolution of the population and housing on flood plains before the year 2000. Despite this, over the recent period between 2000 and 2012, we note an overall continued growth in the Gard's urban areas, with a deceleration of construction in flood-prone areas, which remains modest (Figure 6.4).

Faced with this, the Gard flood risk monitoring center (NOE) estimates that between 2007 and 2013, 10,860 people moved into flood-prone areas: 35% of the Gard population was thus exposed in 2013.

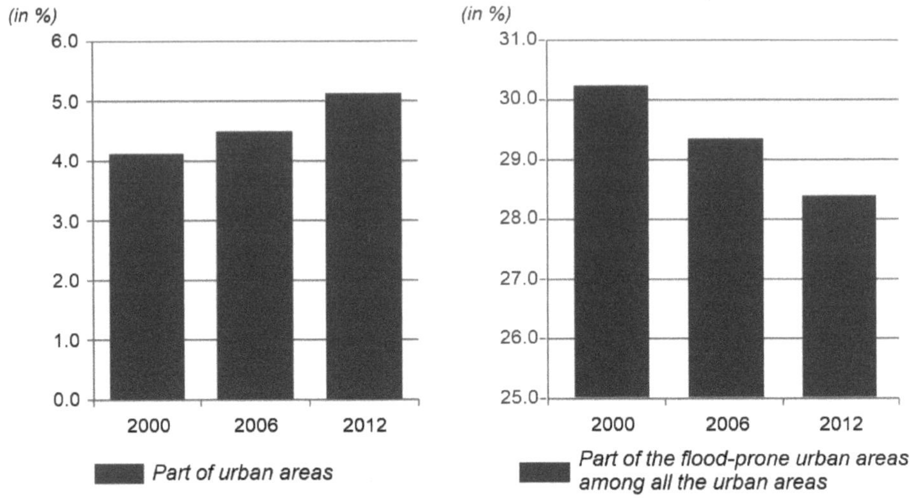

Figure 6.4. *Gard: urbanization faced with the risk of flooding from 2000 to 2012*[5]

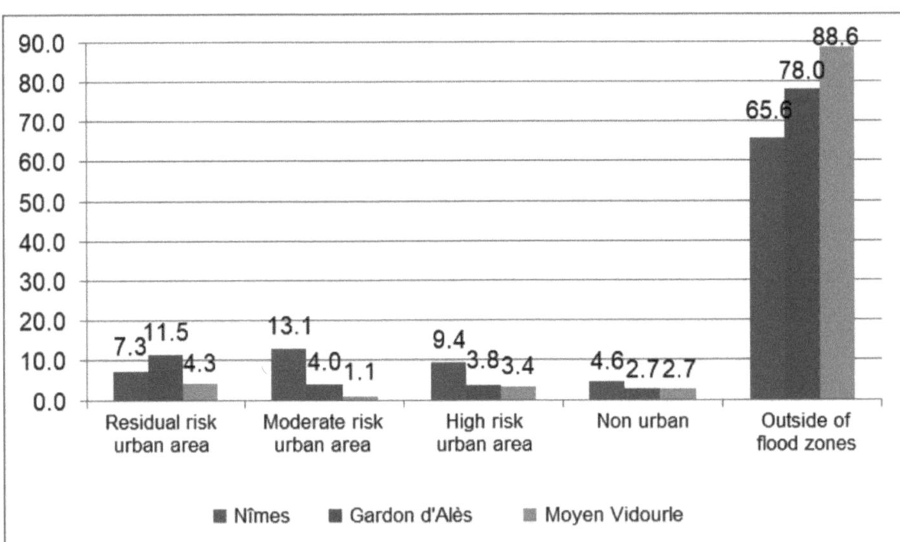

Figure 6.5. *The part (in %) of constructions (all types) between 1979 and 2012 according to PPRI zones. For a color version of this figure, see www.iste.co.uk/lutoff/hydrometeo1.zip*

5 Sources: BD TOPO, IGN, 2012; BD NOE, *Conseil départemental du Gard* (Gard Departmental Council), 2010.

Many Gard residents moved into flood-prone areas. However, nothing suggests that this phenomenon led to new constructions. It is therefore difficult to state that urbanization has continued in flood plains.

To shed light on this reflection, we observed the distribution of new constructions over the 1979–2012 period, following Gard's three PPRI regulatory areas[6] (Figure 6.5). Apart from the areas "outside of flood zones" that group the majority of constructions together, the areas of strong or moderate risk have largely been urbanized; covering more than 20% of construction for Nîmes over that period. Alès is an exception here because, for urbanization in flood-prone areas, residual risk areas are prioritized.

The part covered by constructions in flood-prone areas shows great heterogeneity in space, but also in the type of PPRI regulatory zone, in time and even in the type of construction. The focus we chose for non-urbanized (NU) areas and high-risk urbanized areas (U-F) [7] shows this clearly. Figure 6.6 represents the proportion (%) of new constructions in the two regulatory areas according to the period and the building sites. Certain striking facts should be highlighted:

– in the two regulatory areas, regardless of the PPRI, mainly industrial buildings are found in NU or U-F zones;

– in Nîmes, even over the recent period, a period during which a major flood occurred, urbanization continued strongly in high-risk areas; much like the non-urbanizable areas (Figure 6.6(a)). Residential urbanization is very clearly favored in high-risk areas, with 8–10% of new constructions;

– Alès practically presents a negative urbanization compared to Nîmes. Although the general construction level remains high, in addition to other buildable regulatory zones subject to conditions[8], residential construction essentially appears in NU areas (Figure 6.6(b)). It is the only territory of the three, suggesting an avoidance of U-F areas for residential growth, especially over the recent period;

– on the Sommières territory (PPRI of Moyen Vidourle), although the number of constructions is more modest than in Nîmes or Alès, the proportion of constructions in high-risk areas is spectacular, even for the period during which a catastrophic flood hit the headlines. In these high-risk areas, it is mainly industrial buildings that have appeared, although a resurgence of residential building has been noted in this risk area after the floods and even after the adoption of the PPRI (Figure 6.6(c)).

---

6 Data covering the PPRi of Nîmes, Gardon d'Alès and Moyen Vidourle obtained by dating the buildings of BD Topo from a collection of IGN topographic maps.

7 Areas for which we expected particular preservation.

8 Not represented here for greater readability; these are zones located outside of flood areas of residual or moderate risk.

Despite a relative deceleration of construction in flood-prone areas over the last few years (Figure 6.4), the stakes have nevertheless increased over the period. For example, if a flood similar to that of 1988 reoccurred in Nîmes, then the damage accrued would be of around a billion euros compared to €610 M recorded in 1988 (PUCA, 2013).

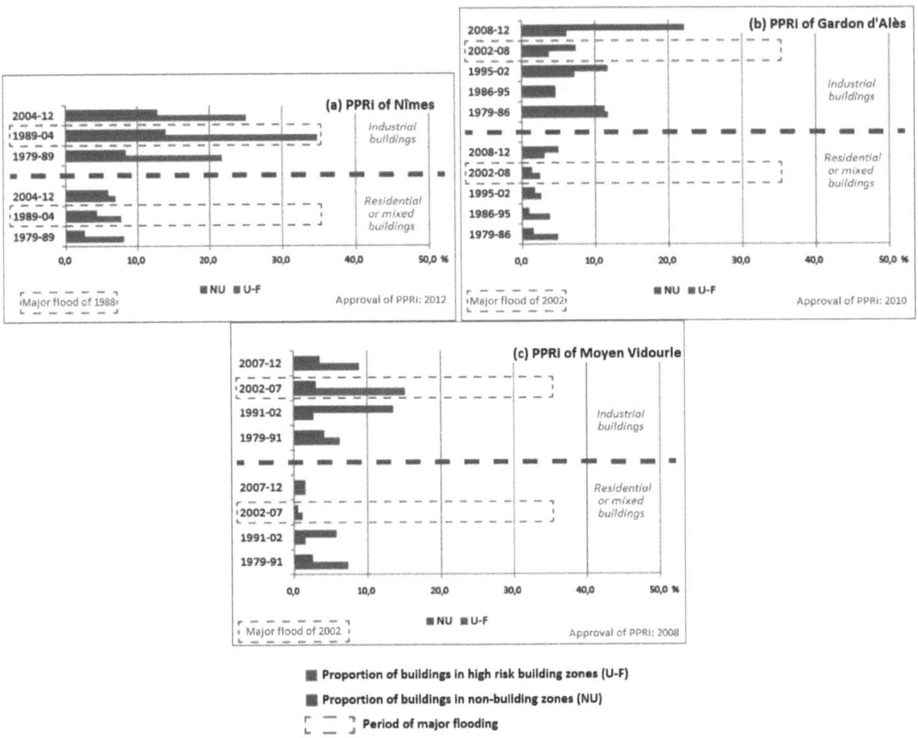

Figure 6.6. *Proportion of the new constructions in non-building (NU) and high-risk to very high-risk (U-F) building zones. For a color version of this figure, see www.iste.co.uk/lutoff/hydrometeo1.zip*

If we follow the example of Gard, a certain number of questions arise relative to the real impact of legal provisions and their legal implications in the protection of goods and people. Considering the number of catastrophic floods that have occurred in Gard, common sense would suggest that we limit construction along waterways, or *a minima*, that building permits systematically have an array of special prescriptions relative to public security. However, a clear deceleration of recent construction in risk zones is noted. Does the absence of a definition for the risk parameters (in the context of a *Plan de Surface Submersible* – (PSS – *submersible surface plans*), a *Plan d'Exposition aux Risques naturels prévisibles* (PER –

predictable natural risk exposure plan), or a PPRI) exonerate the mayor and administration in charge from all of the responsibility of building permits? Moreover, can the case for Force Majeure be made when dramatic events punctuate the history of a region and the unpredictable characteristic of the event can hardly be upheld? The review of jurisprudence on this matter shows that since the early 1990s, the exoneration of public authorities' responsibility has not been systematic.

6.2.1.2. *The responsibility of public authorities and urbanization in a flood-prone area*[9]

Administrative responsibility is engaged if the relevant authority in matters of urban development issues a building permit in a risk area[10]. Article R.111-2 of the French code for urban development constitutes the foundation by which relevant public authorities in matters of urban development can refuse building permits or impose special arrangements if the future construction, through its position, characteristics, importance or set-up near other constructions, is "of a nature to affect public cleanliness or security". Interpreted *a contrario* by the jurisprudence[11], article R 111-2 can serve as a basis for refusing building permits or for attaching special prescriptions to it, notably if the future construction is likely to be exposed to a risk, or even to aggravate a risk to which neighboring sectors are exposed.

Nevertheless, force majeure events can constitute a cause for exoneration from responsibility [COU 93, LAT 45, LED 97]. Force majeure refers to a reasonably inevitable damaging event, as much in its occurrence as in its effects. The event is inevitable because the force at work is considered insurmountable, beyond the will of humans and the means at their disposal. These exceptional circumstances remove all fault in the behavior of the apparent author of the damage, who does not hold any of the responsibility [ANT 92].

To have an exonerating effect, the force majeure event must be *unpredictable*, *irresistible* and *external*. The *external* character of a natural event with regard to the public authority in charge of natural risk prevention or relevant in matters of urban development is barely discussed since it is so obvious – even if the separation between the natural environment and human constructions (that could influence the so called "natural" behavior of a waterway for example) remains debatable in a more general context.

9 For a more detailed historic reconstitution of the construction of the legal framework for flood risk management, see Chapter 4.

10 See one of the first court decisions recognizing the responsibility of the competent authority in matters of urbanism: Lyon administrative court of appeal, December 9, 1992, Mr. and Mrs. Gire, nos 91LY00327–92LY00173.

11 CE, July 13, 2006, French transport, equipment, tourism and sea minister, no. 282937.

Unpredictability refers to an event that cannot reasonably be expected, that we cannot normally predict. Unpredictability is determined in relation to the exceptional character of the observed phenomenon, and notably to its abnormal character (occurrence of an unexpected event or an event of exceptional duration or intensity)[12]. The central element of force majeure, *irresistibility*, refers to an event and its consequences that cannot be avoided or surmounted by the person in question, such that his responsibility is not maintained.

The administrative judge implicitly considers that the phenomenon is not irresistible if *precautions should have been taken and had they been, the damage would not have been caused or would have been less significant*. More often, a harmful event is only irresistible because it was not expected[13]. Unpredictability must therefore be characterized in order to establish irresistibility. The harmful event, although expected, is no less irresistible: an event can be expected and announced, but its magnitude and/or intensity can lead to ineffective prevention or protection systems.

Nevertheless, force majeure is often invoked, but "extremely rarely held up by the judge" according to members of the *Conseil d'Etat* (State Council) themselves [ROU 89][14]. As remarked by Benoît, "force majeure remains a relative notion: it must retreat from the various progresses that allow us to better resist events that may constitute cases of force majeure"[15].

Thus, a new flood occurring in conditions similar to historic floods, all other things equal, may not necessarily have the same character of a force majeure[16]. In this context, force majeure is retained "if the cataclysm is without precedent; the solution is the opposite if the natural phenomenon has already occurred in comparable circumstances and less than a century ago"[17].

---

12 CE, January 23, 1981, City of Vierzon, Rec., p. 28; CAA, Lyon, September 28, 1994, Commune of Bastia, Lexilaser CE CAA. See also Grenoble administrative tribunal, June 2, 1994, Claude Raymond and others: the storm "whose genesis was different to that of previous storms" and the resulting flood "presented, by reason of their exceptional and unpredictable violence and intensity (…) the character of a force majeure event".

13 See Chapter 3 on the challenges of flash flood forecasting.

14 See, for example, CE, March 25, 1988, Territory of New Caledonia and dependent territories, Jollivet and Ludeau, req. no. 56809, cyclonic rains that "due to their predictable character in such a region, do not constitute, despite their extreme violence, a force majeure event".

15 Pr. Benoît, Juris-Classeurs Administratif, ancien fasc. 700, no. 176.

16 CAA Nantes, June 9, 1993; de la Porte des Vaux, req. no. 91NT00682.

17 J. Moreau, Juris-Classeur Administratif, fasc. 700, no. 138. Reference by jurisprudence to a flood of centennial frequency, by default based on older statistics, was thus empirically

With regard to the return period of an event of unpredictable character, the administrative jurisprudence has long required a relatively long time for the statistical return of an event: "An event is considered, by the jurisprudence, as constitutive of force majeure when it occurs at least once per century"[18]. However, the decision of the Lyon Court of Appeal on May 13, 1997[19], in relation to the flooding of Grand-Bornand, marked a turning point in the assessment of a natural event's characterization as a force majeure, by putting an end to this age-old reference: the magistrates deduced the predictability of events based on similar past floods[20]. Despite the long time span between events, the incident is not unpredictable.

Moreover, even when the force majeure is recognized, exoneration from responsibility is not always complete. After review by a judge, if another event contributed to the production of the damage, its author must carry out repairs, relative to their participation, on the detrimental consequences of the wrongful behavior.

In terms of the history of floods in the Gard department, it is therefore difficult for public authorities, and notably relevant authorities in matters of urban development, to invoke force majeure. Certain legal dispositions prevent the development of projects in flood-prone zones, notably through the establishment of risk parameters in 1955 and the establishment of PER in 1982. Moreover, administrative jurisprudence sanctioned authorities who allowed urbanism in zones exposed to a natural risk. With the support of these last two elements, it became possible to limit urban development, particularly along waterways. However, with regard to the observations made in this article, the authorities competent in matters of urban development, such as the State services, did not always prove to be sufficiently reactive to apply the flood risk prevention measures. The occurrence of traumatic events led, at the national level, to the reinforcement of legal measures to

---

adopted without scientific basis. It appears wrong when considering rare cases of known heavier floods. The respective frequencies of floods in Garonne in 1875 and the Seine in 1658 are estimated at 200 years and at least 300 years, and those of floods in the Oder in 1997 are estimated between 500 and 2,000 years (Public report of the French Court of Auditors, 1999).

18 CE, April 4, 1952, French Minister of Public Works against Society of Chais Armagnac, concl. Braibant, *AJDA*, p. 592, 1962, (river flood); CE, January 23, 1981, City of Vierzon cited (floods), no. 13130; CE, March 14, 1986, Commune of Val d'Isère (avalanches), no. 96272; 99725; CE, June 19, 1992, town of Palavas-les-Flots, RDP, p. 68, 1993, (storm); CAA, Bordeaux, April 3, 1995, Consorts Boyer and other (exceptional rains), no. 94BX00378.

19 No. 94LY00923; 94LY01204. *D.*, p. 11, note C., 1998, Shaegis.

20 Six related precedents underwent similar events in the sector.

improve the legal framework for flood risk prevention, and at the local level, the implementation of these legal measures[21].

6.2.2. *Traumatic events and the evolution of risk prevention*

At the national level, no less than 580 flooding events (all types) were reported over the period 1980–2016 alone, in the historic floods database[22]. This impressive number of events is discernible even at the national level. Faced with such catastrophic events, solutions are generally expected from lawmakers for the protection of goods and people, for better public information and more generally, for the adaptation of urban development rules for risks. Two questions are underpinned in these postulates: does the occurrence of events of great magnitude push legislative work in this direction? Do these major events have an impact on the establishment of regulatory texts close to their location?

6.2.2.1. *Legislative evolution in response to major and emblematic events*

To understand the timeframe for the construction of a legal framework for the prevention of natural risks, three main periods are highlighted[23]:

– the first corresponds to the initial developments of the policy until 1995[24];

– the second came about with the promulgation of the February 2, 1995 law;

– the third began with the July 30, 2003 law, notably planning measures specific to the prevention of flood risk.

The compensation law for natural disasters (no. 82-600 from 07/13/1982), established in the risk exposure plans (*Plans d'Exposition aux Risques*, PER), aims to ban new constructions in the most exposed zones to set special conditions for new constructions that are authorized in less exposed zones and to carry out the

21 See Chapter 5.

22 See online database: http://bdhi.fr/appli/web/welcome.

23 We refer the reader to Chapter 5 for a more detailed image of the risk prevention legal framework history.

24 The premises of the policy for the prevention of natural risks can be traced back to the end of the 19th Century, before the law from April 4, 1882 relating to the restoration of mountain land, with the aim of fighting against land movements linked to deforestation. Next, although the 20th Century is studded with legal measures for the prevention of natural risks, with the *Plans de Surfaces Submersibles* (Submersible Surface Plans) in 1935, article R 111-3 of the urban development code in 1955, the *Plans d'Exposition aux Risques* (Risk Exposure Plans) established by the law of July 13, 1982, it was only toward the end of the 20th Century that natural risk prevention policy really hit its stride.

necessary work in order to reduce the vulnerability of the existing construct. PERs were replaced by PPRs in 1995 (law relative to the reinforced protection of the environment).

Since establishing the risk exposure plans (PER), lawmakers on the lookout for prevention methods that are adapted to the various kinds of natural risk, have not ceased to both reinforce legislative prevention measures and to increase the regulatory power of public authorities. The legislative framework for natural risk prevention is therefore reactive and evolves in response to events. Much like environmental law in general, we see that changes often follow on from catastrophes. This is, in fact, a characteristic of environmental law in general, in which laws of risk play a part [SAN 09].

Generally, the occurrence of a catastrophe results in legal modifications to counter the identified shortfalls (Figure 6.7). As such, in the field of risk prevention, major legislative developments appear after a shock of national magnitude. By admission of the authorities themselves, it was the worsening flood risk and notable events that pushed them to strengthen the legislative dispositions.

This is summarized in the foreword of the presentation of the Nîmes PPRi report:

> "The recurrence of catastrophic events over the last ten years, over the whole national Territory, has led the State to strengthen its flood prevention policy. This policy materialized through the establishment of the flood risk prevention plans (PPRi), whose legislative framework is defined in laws no. 95-101 from February 2, 1995, 2003–699 from July 30, 2003 and decrees no. 95-1089 from October 5, 1995 and 2005-3 from January 4, 2005. These are all codified in articles L562-1 and R562-1 and following the Environment code" (Gard prefecture, 2014).

We have also tried to link the most striking events and the main legislative texts for this period. In Figure 6.7, we have presented the major floods of metropolitan France over the period according to the dates of implementation of legislative measures. Thus, we note that the legislative developments of February 2, 1995, follow a series of floods. This all leads to the idea that the two temporal series (that of the adoption of laws and that of floods) are not independent of each other. Although the correlation between two phenomena cannot be formally established, the recurrence of successive events/legislative measures suggests a cause–effect phenomenon. It is probably the magnitude of the damage, and notably the number of victims, that prompted lawmakers to innovate on the matter and research new solutions to respond to the social demand for protection, compensation, etc.

Thus, the law in 1995 gave risk prevention policy a new impetus and made it more coherent [LED 06] through the simplification of measures that were, until then, composed of four different procedures for the same kind of risk[25]. The PPR, whose aim is to outline the areas of risk and plan urban development policy rules and prevention measures in these areas, constituted the turning point around which natural risk prevention policy was orchestrated. Nevertheless, we note that developments in legislative measures appeared with a certain delay after landmark events.

Likewise, the natural risk component in the law from July 30, 2003[26], notably the measures relating to floods, responds to the need for improvement in the field of prevention following floods in the Somme valley between March and June 2001 and floods in Rhône in 2002. These floods alone justify new adapted measures. They complement the general provisions concerning all natural risks. This involves, for example, implementing a reform of meteorological and hydrological surveillance measures and of flood warning systems, or even promoting the amicable acquisition procedure to limit the human stakes in regularly exposed zones, with compensation for the proprietors[27].

Events of large magnitude, whose defining criteria are outlined in Chapter 4, therefore have an impact at the national level and trigger legal, legislative and/or regulatory developments. However, their implementation at the local level is often difficult and delayed.

6.2.2.2. *Weakness in the diffusion of PPRi following events*

We generally note a certain latency period in the initiation of PPR procedures. For example, November 12 and 13, 1999, a few years after the promulgation of the law of February 2, 1995, floods of large magnitude occurred, affecting four departments (Aude, Pyrénées Orientales, Hérault, Tarn) and resulting in 36 victims [LEF 00]. Common sense would suggest that this catastrophic event should lead to a

25 Submersible surface plan for flood risks, risk exposure plan for all natural risks except forest fires, article R 111-3 risk parameter from the urban development code which can be implemented for all risks, forest fire susceptible zone plan for forest fire risk.

26 It should be noted that originally the draft legislation only referred to industrial risks and aimed to bring legislative responses, following the Toulouse accident, to the risk exposure dilemma for buildings near industrial enterprises.

27 Compensation for both the value of the building and the plot itself. The compensation provided by insurance companies with regard to real estate goods does not take into account the plot. Article L 121-17 in the code of insurance refers to the rehabilitation of the plot and not its compensation.

wake-up call for public authorities[28] with regard to the vulnerability of their territory.

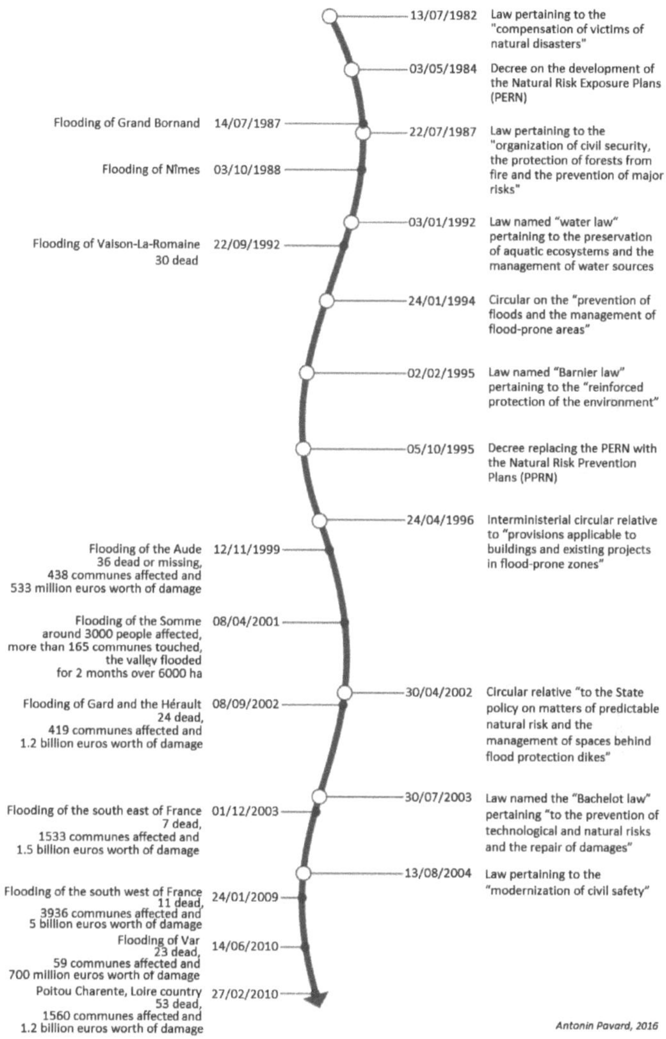

Figure 6.7. *Temporal relationship between large floods and the adoption of legislative texts[29]. For a color version of this figure, see www.iste.co.uk/lutoff/hydrometeo1.zip*

28 Whether in reference to State representatives in the departments (prefects) responsible for implementing the PPR, or communes.

29 According to G. Garry, 2016; www.persee.fr/.

This awareness then triggered the dynamic for prescribing the PPR, protecting the buildings and activities exposed to this kind of risk and limiting, even banning, the development of projects in exposed areas.

Nevertheless, although the feedback and public reports highlighted weaknesses in terms of prevention policy [LEF 00, GAL 01, DAU 99][30], these floods did not have the desired driving effect.

The accumulation of events that occur at a national scale contributes to the development of prevention regulation. However, its establishment at the local level is often only implemented following a major catastrophe in the affected territory. Most often, the occurrence of an exceptionally large event, significant in the history of the territory, drives the launch of a global policy for managing this sort of risk, notably with the introduction of a PPR and the limitation or banning of project development in flood-prone areas. In many cases, it took the 1999 catastrophe in the Aude for the PPR to be applied in impacted areas.

The closer the epicenter of an event to a territory, the more it is felt. Despite this, even if a striking event in a territory pushes for the development of a legislative framework at the national level, the local establishment of a natural risk prevention policy will not, however, easily pass into neighboring territories.

This observation is particularly marked in the Gard department, repeatedly affected by floods. If we take the example of Nîmes and Sommières, two communes among those that are the most impacted by flooding in Gard, we note that they are regularly affected and are the object of multiple decrees recognizing the state of natural disaster. Despite this, it was the floods of the Rhône and Gardon, affecting a large part of the department on September 8 and 9, 2002, that constituted the striking event giving the impulsion to establish the PPR (Figure 6.8)[31]. With this example, we ask ourselves whether it is better to wait for all of the communes of a territory to be directly impacted so that the implementation of regulations is effective.

30 The Dauge report, November 1999, developed before the floods in the same month, also underlines these weaknesses.

31 The 2002 and 2003 floods also had the effect of developing, diffusing or rediffusing the *Atlas des Zones Inondables* (AZI) affecting the Gard department basins: March 15, 2003 for the Gardons basin; January 1, 2001 and July 26, 2004 for the Vidourle basin; January 1, 2001 and February 18, 2003 for the Cèze basin. These AZI were, nevertheless, belatedly developed and diffused in view of the injunction addressed to the prefect in circular no. 94-69 on August 16, 1994 for the mapping of flood-prone areas in the form of an atlas.

Figure 6.8 shows that the developments at the end of the period, resulting with the establishment of the PPRi, undoubtedly come from the multiplication and accumulation of events that have a strong local impact within communities, and even stronger events that have marked everyone's minds (the 2002 event, generalized to the whole of Gard, for example).

However, it is unclear whether it is the accumulation or the generalization of events that leads to the application of laws and the implementation of the PPR. We can hypothesize that the prescription of PPRi is linked to the provisions from the September 5, 2000 decree (appendix I from article A 125-1 of the insurance code).

The implementation of PPRs was influenced by the measures pertaining to insurance premiums[32]. Article A 125-3 of the insurance code thus establishes a mechanism for the modulation of exemptions directly linked to the PPR procedure. In communes without a preventable natural risk prevention plan (PPRN), for a risk with a decree pertaining to the recognition of the status of natural disaster, an exemption is modulated based on the number of decrees taken for the same risk over the five years preceding the date of the last observation:

– first and second decree: application of the exemption;

– third decree: doubling of the applicable exemption;

– fourth decree: tripling of the applicable exemption;

– fifth decree and further decrees: quadrupling of the applicable exemption.

These coercive measures cease to apply when a PPRN which anticipates the risk referred to in the decree, and which supports the observation of a natural disaster status in the affected commune, is prescribed. They take effect again in the absence within a delay of four years of approval for the aforementioned plan, from the effective date of the decree of the recommended plan.

Unfortunately, the implementation procedures of a PPR started late in the Gard department: following the occurrence of seven floods classified as "natural disasters" (since the establishment of the natural disaster guarantee in 1982) in the commune of Alès; after six floods of this kind occurred in the commune of Bagnols-sur-Cèze; after 14 decrees of natural disaster in the commune of Nîmes. However, a risk parameter based on article R 111-3 of the urban development code

32 And by the incitement of the decree of September 5, 2000, modified by the decree of September 10, 2003.

was established after seven floods classified as natural disasters occurred after 1982 and after seven floods occurred in Sommières (Figure 6.7). Moreover, despite the measures in Appendix I of article A 125-1 of the insurance code, which includes the modulation of insurance exemption in the absence of approval within four years of the date of decree of PPR prescription, the delay between this date of prescription and the date of approval was very long: respectively, nine years and three months for Alès, four years and four months for Bagnols-sur-Cèze[33]. Only the PPR of Nîmes and Sommières were approved within a delay of less than four years (three years and four months, and three years and seven months, respectively).

6.3. Construction in flood-prone areas: a heterogeneous consideration of the legislative constraints of territory management

The legislative arsenal for promoting protection against floods has been enriched over time. However, the legislative constraints are not applied homogeneously over the whole Gard department. These constraints do not hold the same weighting for urban development in all flood-prone areas. Two main reasons can be put forward to answer the question of "why did we continue to build in flood-prone areas when even the PPRi oppose the PLU?": a differentiation in the application methods of legislative measures, and a differentiated reactivity in the awareness of flood risk in urban development documents.

The pursuit of urbanization in flood-prone areas plausibly arises from the multiplication of other contextual factors and other issues (need for accommodation, economic needs, etc.) in legal implementation. On the one hand, social demand remains strong on the matter of accommodation. On the other hand, economic activity also requires the construction of new buildings. Furthermore, many communes do not have much free space in non-flood-prone areas. As such, faced with the imperative of responding to the different social injunctions, community leaders are tempted to opt for a positive economic and social outcome, leaving the protection measures for goods and people more or less in the background.

Concretely, on the ground, this translates into great difficulty in obtaining territorial legal responses for floods alone and a slowness in taking flood risk into account in urban development documents.

33 During this delay, a flood occurred in Alès and three occurred in Bagnols-sur-Cèze.

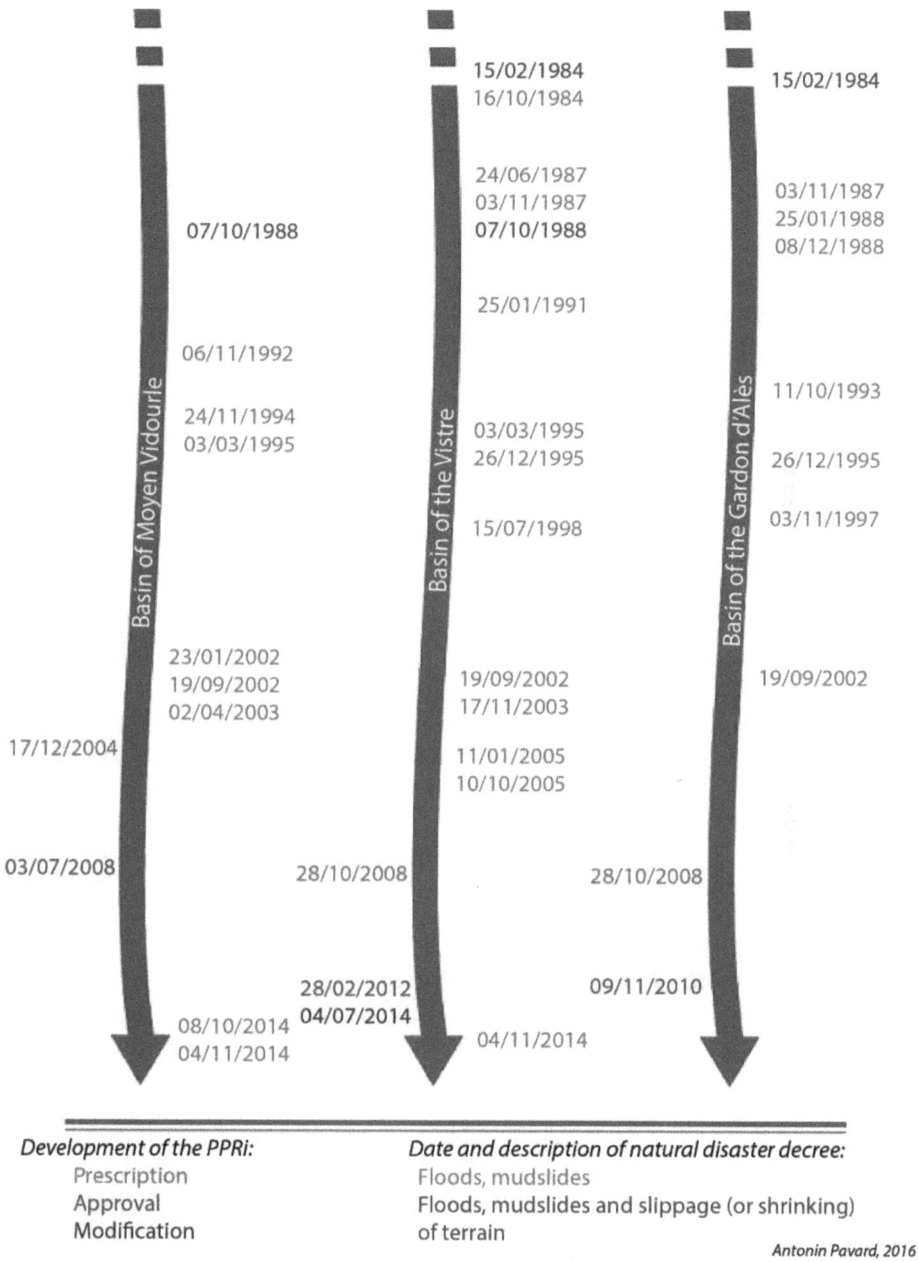

Figure 6.8. *Natural disasters due to floods and the development of the Gardon PPRi in Alès, Moyen Vidourle and Nîmes. For a color version of this figure, see www.iste.co.uk/lutoff/hydrometeo1.zip*

6.3.1. *A variable implementation of legislative measures from one territory to another*

The PPRi was established by the 1995 law pertaining to the organization of civil security, the protection of forests against fire and the prevention of major risks. Based on this law, urban development documents had to take risks into account to determine the constructability of sectors in terms of zoning and regulations.

Although the general framework is given at the national level, documents relating to the PPRi are difficult to compare, in terms of form as much as in content. Whether in terms of mapping, regulation or even breakdown into zones, many elements differ from one PPRi to another, even for adjoining territories [MOU 13, DOU 11]. Noticeable differences can also appear within different versions of the same PPRi, as seen in Nîmes, in which certain commune areas, although locked in flood-prone areas, are excluded from the regulation in the first version of the 2012 new generation of PPRi, before being reintegrated in the 2014 reviewed version (such as, for example, in the west of Mas de Possac). However, over the last few years, we nevertheless note a willingness to increasingly harmonize the PPRi.

In the context of PPRi establishment, regulatory zoning spatially expresses the hazards. It is drawn up by taking into account past events and the stakes in the observed territory. Without touching on the problem of the level of variability in the zoning and mapping details themselves, the PPRi are most often divided into three or four areas which refer to a specific level of hazard: residual, medium, high and very high, referring to urbanized areas or areas susceptible to becoming urbanized, but also areas without dikes and flood areas. Although all PPRi developers seem to respect this nomenclature, disparities sometimes also hide in the interpretation of these classifications.

Zoning established based on levels of hazard serves as a basis for defining actions and developments authorized on the territory. Most often, the PPRi regulations for the banning or authorization of a development rely on a general nomenclature, common to all PPRi. However, local specificities can appear depending on each territory's own stakes and according to the interpretation given by the public powers in charge of establishing them. Thus, by comparing the regulatory areas of Moyen Vidourle (Sommières region) and of Gardon d'Alès, certain divergences can be observed: for example, in Moyen Vidourle, there is no restriction made for the construction of traveler reception sites, whereas it is forbidden in flood-prone areas of the Gardon d'Alès (Table 6.2). This is an element that lifts the curtain on the potential manipulation of regulations to make less popular political measures more easily accepted.

	Moyen Vidourle						Gardon d'Alès					
	Urbanized			Non-urbanized			Urbanized			Non-urbanized		
Hazard level	H	M	R	H	M	R	H	M	R	H	M	R
New constructions	☒			☒	☒	☒	☒			☒	☒	☒
New constructions or extensions of existing constructions located less than 10 meters from a registered watercourse or a registered moat				☒	☒	☒		☒	☒	☒	☒	☒
Development and extension or capacity increase of traveler reception sites							☒	☒	☒	☒	☒	☒
Development of a cemetery							☒					
Reconstruction of buildings affected by a flood	☒	☒	☒	☒	☒	☒	☒	☒		☒	☒	☒
Creation of openings beneath the reference slopes (PHE + 0.30m) if these are not equipped with cofferdams of at least 0.5m in height	☒			☒	☒							
Greenhouses	☒						☒			☒	☒	
Deposit and storage of dangerous products or pollutants	☒	☒		☒	☒		☒	☒		☒	☒	☒
Equipment of public interest										☒	☒	

Table 6.2. *Restriction level according to the zoning of two PPRi: Moyen Virdoule and Gardon d'Alès (☒ – restricted actions; hazard – H(igh) M(edium) R(esidual))*

With this example, we clearly see how varied the measures taken for flood protection are. Firstly, restrictions are particularly effective: it is very present, regardless of the level of hazard, in urbanized or non-urbanized areas and for all reconstruction of buildings affected during flooding. The implementation of the PPRi in such cases shows a willingness to reduce particularly exposed buildings. The same applies to the storage of dangerous materials that can be threatened by floods. For new constructions, the rules are a little more flexible.

In non-urbanized areas, social demand is less strong and restriction is easier. This is not the case in urbanized areas where the pressure due to the need for accommodation and buildings for activities has led to the search for a compromise, such as putting new constructions *in extenso* under constraints in the case of medium or residual hazard. For other projects, like the two examples in Table 6.2, the levels of restriction are much more differentiated, thus marking more obvious local negotiations.

6.3.2. *Flood risk is taken into account belatedly in urban development documents*

Since the law in 1987[34], measures have been taken to reinforce the awareness of natural risks, and particularly the risk of flooding, in urban development documents.

Thus, the SCOT must determine the urban development approach by taking into account the natural risks, and since the ruling on September 23, 2015, this must especially be compatible with the flood management objectives defined in the *Plans de Gestion des Risques d'Inondation* (PGRI – flood risk management plans) and thus the fundamental approaches are defined[35]. Through the demarcation of building or non-building areas, the PLU also contributes to the prevention of natural risks by controlling some of the stakes in exposed areas and by potentially defining specific recommendations[36]. Like the SCOT, the PLU must also be in agreement with the requirements defined by the PPRi as well as the fundamental approaches

34 Law no. 87-565 of July 22, 1987, pertaining to the organization of civil security, the protection of forests from fire and the prevention of major risks.

35 Article L 131-1 of the urban development code. When a flood risk management plan is approved after the endorsement of a SCOT, the latter must, if necessary, be rendered compatible with this plan and its approaches within a delay of three years. In this case, the SCOT do not need to be compatible with the fundamental approaches relating to the prevention of floods defined by development and water management master plans (article L 122-1-13 of the urban development code).

36 However, the PLU cannot issue technical construction rules, nor intervene on existing constructions.

and measures of these plans, once they are approved[37]. The communes therefore do have competencies in the field of urban planning and risk, especially since the prefect conducts a legality audit of the urban development documents, to check their conformity to the different applicable requirements.

The obligation of taking flood risk into account in urban development documents is clearly articulated today. This was not the case before the PPRi were imposed onto other urban development rules. Thus, despite the recurrence of floods in the observed department, the urban development documents of Gard's communes show that the *Plans d'Occupation des Sols* (POS – land occupation plans), which preceded the PLU, did not clearly mention the risk of flooding, nor did they precisely outline the flood-prone areas.

	Alès	Bagnols-sur-Cèze	Nîmes	Sommières
POS	05/08/1979	02/15/1985	02/04/1983	08/09/1983
PLU	06/24/2013	02/13/2006	03/01/2004	02/26/2008

Table 6.3. *History of urban development documents in the studied area*

As seen in Table 6.3, the first urban development documents (POS) in the studied area were developed before 1987. In consequence, these first measures taken in urban development had no legal obligation to take into account the risk of flooding. However, despite the absence of legal constraints and considering the numerous floods that occurred in the department, we could have expected a limited occupancy of flood-prone areas, with the aim of protecting people and goods, especially since the AZI were available for certain catchments of the department as of 2001 and the flood-prone areas were therefore known.

The risk of flooding is certainly mentioned in the POS that we have studied, but with various degrees of precision and awareness. This had variable consequences on the development of urbanization in flood-prone areas. For example, the POS of Alès was promulgated in May 1979. Its regulation contains only a requirement

37 When a flood risk management plan is approved after the endorsement of a PLU, the latter must, if necessary, be rendered compatible with these elements within a delay of three years. In this case, the PLU does not need to be compatible with the fundamental approaches relating to the prevention of floods defined by development and water management master plans (article L 123-1-10 of the urban development code).

concerning the location of constructions, with the obligation of respecting a certain distance (between 8 and 15 m depending on the sectors) between the construction and the crest level of dikes. The extract from Alès commune's POS (Figure 6.9) does not clearly show the flood-prone areas, and certain areas were at the time defined as buildable. At the end of the 1970s, the POS zoning showed sectors that were not yet built up. However, in 2010, these same areas were built up despite being in high-risk areas at the PPRi level (Figure 6.10). These constructions were therefore developed, in sectors exposed to a flood risk, between the beginning of the 1980s and the promulgation of the PPRi. There is therefore a great paradox here, since this period represents the first deployment phase of the national natural risk prevention policy following the promulgation of the law on July 13, 1982, which notably established the PER.

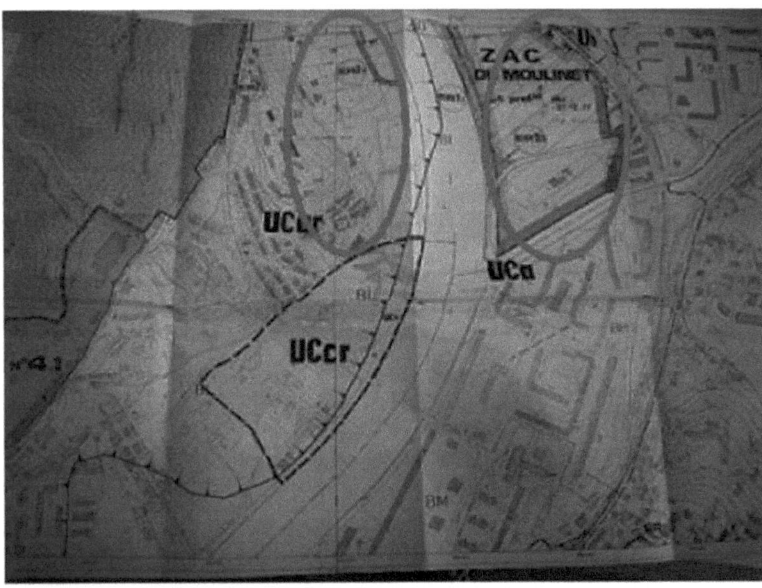

Examples of non-urbanized or weakly urbanized areas at the time of the establishment of the POS (1979)

Figure 6.9. *Extract from Alès commune's POS (1979). For a color version of this figure, see www.iste.co.uk/lutoff/hydrometeo1.zip*

For the commune of Sommières, the first POS (dating back to 1979) also took into consideration the risk of flooding. However, according to the PPRi mapping, an important part of the historical center of the commune bordering Gardon is in a flood-prone area. Taking this situation into account, the POS regulation prohibited the use of ground floods for accommodation in buildings located close to Gardon.

It was also prohibited to build extensions for buildings located in flood-prone areas. The POS regulations state that proprietors had to implement these prevention and protection measures which were needed to protect their buildings, although without stating the potential measures to be implemented.

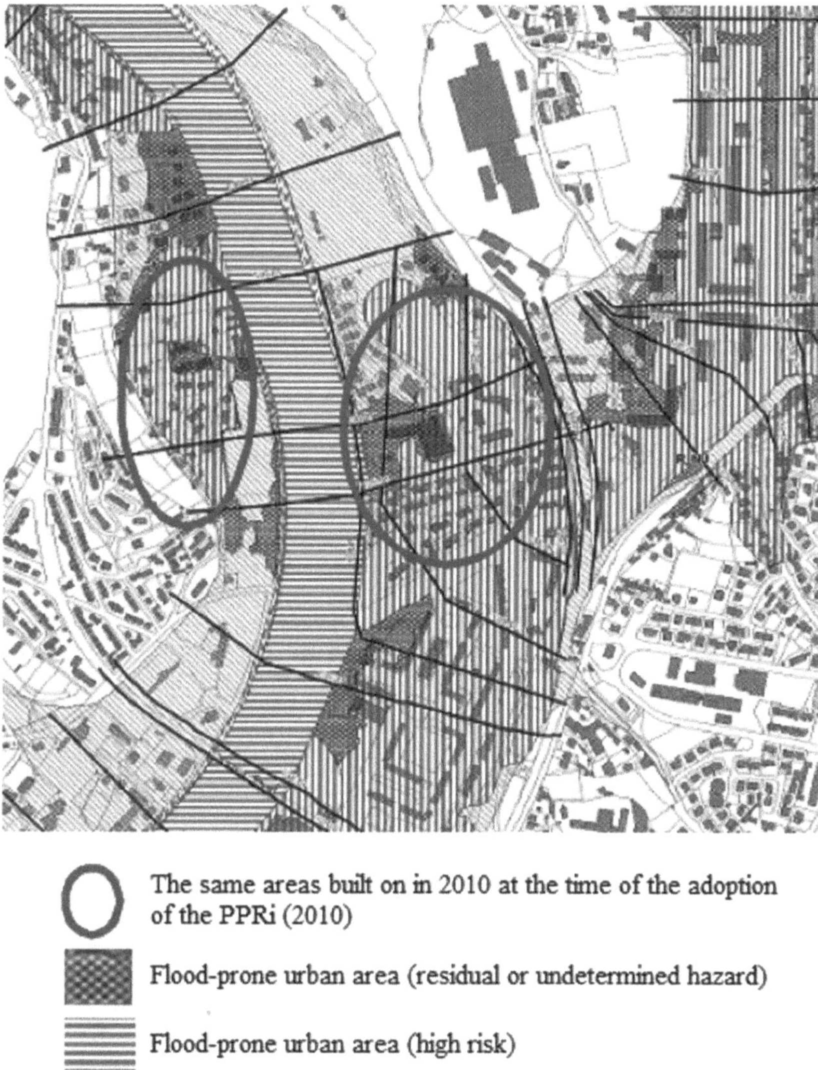

Figure 6.10. *Extract from the PPRi of Alès commune (2010). For a color version of this figure, see www.iste.co.uk/lutoff/hydrometeo1.zip*

Despite these elements of knowledge on flooding, we make the same observation based on a comparison of the POS (1983) and the PPRi (2008) mapping documents: over 15 years, many buildings appeared in areas of high or moderate hazard, despite knowledge of these hazards (Figures 6.11 and 6.12). Moreover, we also note a smaller influence of the unbuildable areas in the POS compared to those outlined in the more recent PPRi.

As suggested through these two examples, it is ultimately only under the impulsion of the PPRN that flood risk became explicitly integrated in urban development documents. The reform made in the law of December 13, 2000, together with the ensuing laws, forced communes to develop a PLU to replace their old POS, and this was an opportunity to determine the constructability of areas by reassessing the PPR flood zoning (which, as a reminder, were taxable in the first POS).

The reality remains that the modalities of flood risk integration leaning on the PPRN are variable from one commune to another. As such, certain PLU explicitly state the flood PPR areas and show through the area denomination that the flood areas and regulations of the PLU reproduce, according to areas, the measures and constraints of the PPR. Other PLU are satisfied with falling back on the PPRi regulations by indicating that the applicable measures are those that are the most restrictive and that the zoning does not clearly outline the flood risk, or, in other words, the PLU zoning is not completely coherent with that of the PPRi.

The Nîmes PLU, from March 1, 2001 and whose latest revision date was in September 2013, clearly illustrates this situation. The PPRi flood areas (all categories) are in the areas of urbanization (AU) of the Nîmes PLU. These spaces, destined for urbanization projects, therefore risk contradicting the PPRi regulatory recommendations. The high-risk areas are unbuildable in terms of the PPRi but nevertheless overlap with several AU areas of the PLU. In order to avoid giving up these urban expansion areas, needed to face the strong social demand within a tense urban context, the commune proceeded to carry out work to move these areas for urbanization "out of water". However, we now ask ourselves whether there is a risk of flood exacerbation in the neighboring sectors in which, for example, rises in water could be capped (retained, etc.) in order to spare the denser urban expansion areas. Similarly, a large part of the *Zones d'Aménagement Concertée* (ZAC – concerted development areas), which notably welcome artisanal or industrial activities, find themselves in flood-prone areas, as defined in the PPRi.

The differences between the PLU and PPRi are not limited to a simple superimposition of different areas.

Figure 6.11. *Extract from the POS of Sommières commune (1983). For a color version of this figure, see www.iste.co.uk/lutoff/hydrometeo1.zip*

Before jumping to rapid conclusions, we should ensure that the regulatory areas correspond to the reality. The State has the power to control the legality of local urban development regulations. However, in its legal note pertaining to the eighth modification of the Nîmes PLU, the State highlighted this kind of gap[38]:

> "In reference to the items of the regulation affecting activities in flood-prone areas, in section VIII AUd, located in the non-urban very high hazardous area of the Nîmes PPRi, no construction can be accepted – even within a limit of 100m^2 floor surface. If such construction is necessary, it must be done in a nearby areas where authorized. Moreover, in section NT01 corresponding to camping in Bastide, the latter is inappropriately equated to an 'area of public equipment for sports and leisure' introducing ambiguity and incorrectly suggesting that this kind of activity authorizes limited construction in flood-prone areas. However, camping is in no case equitable to 'public equipment for sports and leisure'. Indeed, in the approved PPRi, camping denotes accommodation where only limited

38 State legal note CS214286 from July 17, 2012, DDTM30.

extension of existing construction is authorized. The formulation must be corrected in the regulation".

In a context of urban expansion, the legal arsenal allows local authorities to establish prevention and protection measures against the risk of flooding. The development of protection plans and their application therefore result from a compromise between the prevention allowed by the law and the other social demands to which the public powers must also respond. This "conciliation" takes on multiple forms that are intricately linked to local conditions.

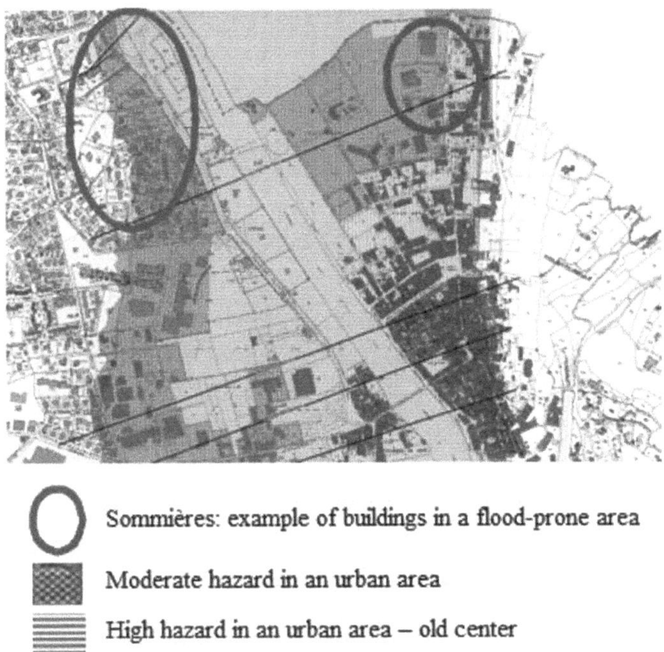

Figure 6.12. *Extract from the PPRi of Sommières commune (2008). For a color version of this figure, see www.iste.co.uk/lutoff/hydrometeo1.zip*

6.3.3. *Measures for reducing vulnerability adapted to the existing stakes*

To what extent can we conserve urban protection and development in a context of high social demand? By bringing the Nîmes PLU and PPRi of the same area together, we observe that there is a relative leeway that can be qualified as a functional clearance between the establishment of urban development documents and knowledge of the dangerous flood areas underlined by the PPRi.

This relative latitude is based on individual developments targeted when important stakes coincide with highly hazardous areas. These individual developments enable the coexistence of a tense urban context with a presence of risk that is strongly reminiscent of the PPRi. These adapted measures allow for some continuation of urban development projects.

For existing constructions, the constraints imposed are more or less heavy, more or less coercive. They involve, for example, the creation of a refuge space in case of flooding, accessible from the inside with an access to the outside, with ballast, anchored tanks and pollutants, a differentiation between the parts out of water and those under the reference line on the electrical network, and even the replacement of material that is sensitive to water with water-resistant material.

In terms of existing activities, the PPRi often anticipate restrictions in extensions for camping, caravanning and recreation grounds, for the creation of a refuge area and the establishment of a safeguarding plan. It can also be mandatory to record the positions of pools, reservoirs and manholes. In high-risk areas, the storage of pollutants and dangerous materials that have been either entirely or largely damaged by a flood, the extension of a perimeter or increase in the number of camping grounds and existing residential recreation parks, may be prohibited.

However, when studying the different PPRi in effect in the Gard department, we note that there are fewer measures concerning existing constructions and activities than there are imposed on future constructions and activities. This is easily understandable: it is easier to impose recommendations to be respected on future constructions and activities, as these will be included from the very conception of a project. The implementation of measures to reduce the vulnerability of existing goods faces several limitations. In studies dedicated to urban renewal and "zero damage"[39] accommodation, the CEPRI (*Centre Européen de Prévention du Risque d'Inondation* – European center for the prevention of flood risk) proposes implementation limits and conditions for flood adaptation strategies in constructions (avoid, resist, divest). None of these are perfectly adapted for floods with high flow velocity. In these areas, only extremely heavy and expensive construction techniques have the potential to be effective, but their cost is too high to be supported with respect to the frequency of flooding [CEP 09]. In other cases, as soon as the construction or activity has been established, the intervention modalities for reducing their vulnerability become limited, for technical and/or financial reasons. The technical reason is linked to the fact that once established, measures that could

39 CEPRI, Comment saisir les opérations de renouvellement urbain pour réduire la vulnérabilité des territoires inondables face au risque d'inondation – Principes techniques d'aménagement, Report, February 2015. CEPRI, Un logement "zéro dommage" face au risque d'inondation est-il possible ?, November 2009.

have been effective in decreasing vulnerability become very limited, with respect to the configuration of the location, the structure, the building, etc. For example, street-level constructions, popular between the 1960s and 1980s [CEP 09], are built on foundations that cannot support the addition of an extra level. However, the PPR sometimes impose the construction of a refuge space on another level. What should be done in these situations? The only option is the creation of an opening on the roof, easily accessible from the inside and providing access to the outside. In flood-prone areas where water levels are above 1.75 m, it is difficult to elevate electric, heating and boiler equipment. Street-level buildings in flood-prone areas present a case for which few building improvement measures aiming to decrease its vulnerability can be planned; although perhaps it is not just the creation of refuge spaces in certain cases[40].

Financial reasons are an additional difficulty linked to the establishment of protection measures: the spending threshold for prevention work proposed in article R 562-5 of the environment Code, concerning the total amount for the works imposed by a PPR regulation, cannot be of more than 10% of the market value of the good itself.

Despite all their virtues, the individual "compensatory" measures could be all the more effective if they were in a more general framework, in which all of the collective protection measures could be taken [CEP 09].

6.4. Conclusion: the relative effectiveness of legal limits and measures

Faced with the risk of flooding, a set of regulatory measures has been adopted. However, their application was not without challenges, at least until the mid-1990s. The establishment of the *Plans de Prévention des Risques Naturels Prévisibles* (1995, predictable natural risk prevention plans) is often considered as a decisive step toward taking risks into account in France [LED 06]. However, it must be noted that urbanization in risk areas continues, as shown in section 6.2 (notably Figures 6.4 and 6.5). Social demand appears to have pushed the elected leaders to prioritize the needs in accommodation and activity to the detriment of the protection of individuals and goods.

The relative abatement of flooding between 1958 and 1988 was accompanied by a notable demographic growth and densification of urban areas in Gard. Although

40 "The obligation to *a minima* make a refuge area in the existing building only concerns street-level habitations with the potential of receiving more than one meter of water. For this reason, the altimetric level of the first habitable floor is then taken into account" (vendee.gouv.fr).

harmonization of data does not allow for a precise evaluation of the population in flood-prone areas, we nevertheless note a very relative decline in exposed parts of urbanized areas. It is no less remarkable that 35% of the population of Gard is in flood-prone areas. Slowing down of constructions in flood-prone areas is a very recent phenomenon: the catastrophic events that occurred over the last few years were taken into account with difficulty.

The evolution of risk prevention was a long process whose key stages very often occurred after events of major, emblematic or unrivalled magnitude. Without establishing a mechanical link, the confrontation between the time of the striking events themselves and the work of lawmakers on the matter shows a certain temporal coincidence. However, this clash also shows a certain inertia; an inertia that we also find in the local initiation of PPR procedures. This delayed implementation is further reinforced by the weak spatial diffusion of decisions taken in territories that are, nevertheless, neighbors. The increased construction and renovation costs imposed through the application of new regulations, as well as the urbanization freeze in certain areas, generate a real difficulty in the development of the territories that these actors must deal with. The problems induced by taking risks and their feasible solutions into account depend on each territory's context: it is therefore not surprising that the spatial diffusion of applicable regulations is not very active.

The problem is further complicated by the heterogeneous application of legal constraints. The imposable nature of the PPRi in the PLU is not applied in the same way everywhere. There remains a large flexibility margin for the local authorities responsible for the development of their catchment areas.

The ambiguous application of rules for risk therefore constitutes a brake in the rigorous establishment of prevention measures. This explains, in part, the tardiness in taking flood risk into account in urban development projects, shedding light on the pursuit of construction in flood-prone areas. The environment code does not formally anticipate any obligation of the State in the establishment of PPR[41].

---

41 The European *Directive Cadre Inondation* (DCI – Flood framework directive) is partially imposed on the national regulatory framework by demanding that States establish, among others, a "flood risk management plan that is coordinated at the hydrographic district or coastal level". Although these plans must mandatorily include a situational analysis (preliminary evaluation of flood risk and flood mapping), they remain relatively motivational since they "may also include encouragements to implement durable land use methods, improvements in water retention, as well as the controlled flooding of certain areas in case of an episode of flooding" (Directive 2007/60/CE of the European Parliament and Council from October 23, 2007 pertaining to the evaluation and management of flood risk).

Their implementation therefore depends on the assessment of the situation by the prefect. Moreover, to our knowledge, the State has never been condemned by the administrative court for shortfalls in the implementation of the PPR, whether due to their inaction or the insufficient nature of the prevention measures. On the other hand, the issuance of authorization for urban development on a plot exposed to a natural risk generates liability due to the lack of a relevant authority in matters of urban development, which is most often the mayor [DOU 11].

Risk prevention at the local level therefore represents a subject that is both delicate and complex, in which there is often a balancing of the risk itself and the need to respond to other issues in the territory. This difficulty is all the more important if potential reconsiderations of mandates made by the decision-makers following such and such unpopular decisions are taken into consideration. This situation, nevertheless, remains all the more surprising considering that there are not many exonerations of public authority responsibility: the case of force majeure, often invoked, is only rarely retained by the court, notably because unpredictability can not be admitted when the events are repeated in ways similar to known historic events.

Finally, in the face of all the difficulties linked to the management of constructions in communes, indispensable to the local economy and projects, this study highlights that with the willingness to pursue urbanization, insurance provides powerful leverage against policy: the increase in insurance premiums manifestly influences decisions in terms of protection and the security stakes of goods and people, faced with a need for urbanization which appears inevitable.

6.5. Bibliography

[ANT 92] ANTONMATTEI P.-H., "Contribution à l'étude de la force majeure", *Bibliothèque de droit privé*, vol. 220, Paris, 1992.

[CEP 09] CEPRI, *Un logement 'zéro dommage' face au risque d'inondation est-il possible ?*, November 2009.

[COM 14a] COMMISSARIAT GENERAL AU DEVELOPPEMENT DURABLE, L'environnement en France, Paris, 2014.

[COM 14b] COMMISSARIAT GENERAL AU DEVELOPPEMENT DURABLE, Indicateurs de développement durable territoriaux – la population exposée à des risques d'inondation par cours d'eau, Paris, 2014.

[CON 16] CONSEIL DEPARTEMENTAL DU GARD, Le Gard Noé, Nîmes, 2016.

[COU 93] COUZINET J.F., "Cas de force majeure et cas fortuit : causes d'exonération de la responsabilité administrative", *Revue du Droit Public*, no. 5, pp. 1385–1426, 1993.

[DAU 99] DAUGE Y., Les politiques publiques de prévention des inondations, Assemblée Nationale, Paris, 1999.

[DEG 02] DEGARDIN F., "Urbanisation et inondations : de l'opposition à la réconciliation", *Bulletin de l'association de géographes français*, vol. 79, no. 1, pp. 91–103, 2002.

[DOU 01] DOUVINET J., DEFOSSEZ S., ANSELLE A. *et al.*, "Les maires face aux plans de prévention du risque inondation (PPRi)", *L'espace géographique*, vol. 40, pp. 31–46, 2001.

[GAL 01] GALLEY R., FLEURY J., Les causes des inondations répétitives ou exceptionnelles et sur les conséquences des intempéries afin d'établir les responsabilités, d'évaluer les coûts ainsi que la pertinence des outils de prévention, d'alerte et d'indemnisation, Assemblée Nationale, Paris, 2001.

[LAT 45] LATOUNERIE R., "De la faute et du risque à propos des dommages causés par les travaux publics", *Revue du Droit Public*, Paris, 1945.

[LED 97] LEDUC F., "Catastrophe naturelle et force majeure", *Revue Générale du Droit des Assurances*, vol. 2, no. 5, 1997.

[LED 06] LEDOUX B., *La gestion du risque inondation*, Lavoisier Tec & Doc, Paris, 2006.

[LEF 00] LEFROU C., MARTIN X., LABARTHE J.P. *et al.*, Les crues des 12, 13 et 14 novembre 1999 dans les départements de l'Aude, de l'Hérault, des Pyrénées-Orientales et du Tarn, Inspection générale de l'environnement, Paris, 2000.

[MOU 13] MOULLIN E., DEROUBAIX J.F., HUBERT G., "La constructibilité des zones inondables ou la négociation territoriale de la règle PPRi", *Géocarrefour*, vol. 88, no. 3, pp. 173–182, 2013.

[PRE 14] PREFECTURE DU GARD, Communes de Nîmes – Rapport de présentation du PPRi, Ministère de l'Intérieur, Services de l'État dans le Gard, Nîmes, 2014.

[ROU 89] ROUGEVIN-BAVILLE M., DENOIX DE SAINT MARC R., LABETOULLE D., *Leçons de droit administratif*, Hachette, Paris, 1989.

[SAN 09] SANSEVERINO-GODFRIN V., L'évolution du cadre juridique des risques naturels et industriels, HDR, Université Nice Sophia Antipolis, Nice, 2009.

[VIN 10] VINET F., *Le risque inondation diagnostic et gestion*, Lavoisier Tec & Doc, Paris, 2010.

7

The Influence of Residential Migrations on the Evolution of Populations Exposed to a Flood Risk

7.1. Introduction

Chapter 6 demonstrated the difficulties of implementing risk management regulations in the Gard department, often affected by floods. One of the factors explaining these difficulties is the strong social demand for the region over the last few decades. In this chapter, we aim to explore in more detail the role played by residential migrations in the evolution of the exposure of the Gard's population to floods.

The availability of harmonized data since 1968 allows the (re)evaluation of population dynamics within the department of Gard, marked by a notable demographic growth, essentially pushed by the migratory balance. However, amongst the 353 communes that make up this territory, not all have followed the same trajectory. The mining industry crisis, for example, severely affected many communes in the Cévennes region, whereas the development of tourism on the coast triggered a marked population growth.

Migrations thus triggered a population growth in flood-prone areas, although this trend has decreased in recent years. It is therefore of interest to identify the different factors that have favored these residential migrations since 1968, and to analyze whether or not they have increased the vulnerability of the population to the risk of flooding.

To identify them, estimate their importance and define their spatial extent, we decided to combine a literature review with modeling in a systemic way. This

---

Chapter written by Edwige DUBOS-PAILLARD, Antonin PAVARD and Joël BOULIER.

review guided the construction of a model whose outcome fed back into the literature review. The different simulations improved our knowledge of residential dynamics from known factors, affecting a large part of the department, and gradually allowed us to target more local factors with specific dynamics.

These demographic dynamics were then compared with the evolution, since 1968, of the population in the flood-prone areas of 33 communes in Gard. The analysis revealed the role of local politics and the effects of legal flood risk regulation improvements over the 45 observed years.

7.2. Gard: a dynamic department in which more than one out of three people is exposed to a flood risk

Gard experienced an important demographic growth from the early 1960s. Over 50 years, the population multiplied by 1.7, reaching 725,600 inhabitants in 2012. Despite a deceleration between 1968 and 1975, since the early 1960s the department has undergone an average annual population growth of 1%, making it a dynamic space in metropolitan France (Figure 7.1) and adjacent departments. Only Hérault (+1.5% per year) and Vaucluse (+1.2% per year) have had a greater demographic growth.

This evolution is due to the migratory balance which, alone, explains the large majority of the observed population dynamics within the communes ($R^2 = 0.99$[1] for the 1962–1968 period). Although this contribution slightly decreases with time, it remains essential in explaining the demographic evolution noted during the last intercensal period studied (R^2=0.86 for 2007–2012).

Similar to other departments of the Mediterranean region, Gard fully benefitted from phenomena of heliotropism and coastal attractiveness. At the same time, the urbanization process that marked the second half of the 20th Century favored the concentration of populations in major urban centers, in parallel with the urban sprawl toward rural peripheries. In 2014, 55% of the Gard population lived in urban centers, 18% in suburban peripheries; the respective portions were of 49% and less than 1% in 1968 (Figure 7.2). The Gard urban network is, first and foremost, marked by the weight of the Nîmes agglomeration, located in the south-east of the department, which holds a quarter of the population and more than one-third of the department's jobs. Alès is 40 km away. It holds two times fewer inhabitants and Bagnols-sur-Cèze, in the north east, 12 times fewer. The specificity of the department lies in the importance of its multicentric municipalities: 23% of the population lives in communes under the

1 R^2 corresponds to the coefficient of determination that indicates the variability in response variable y (population dynamics) that can be attributed to explanatory variable x (apparent migratory balance). The coefficient of determination is the square of the linear correlation coefficient).

influence of several urban centers. Although Nîmes extends its influence over the center and south east of the department, it competes increasingly with Avignon in the east, Monpellier and Lunel in the south west, and Alès in the north west [BUF 16].

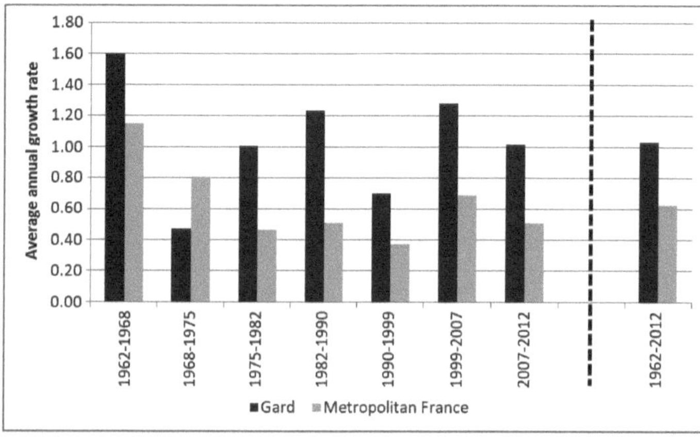

Figure 7.1. *Evolution of the Gard population between 1962 and 2012 (sources: INSEE, RGP 1962–2012)*

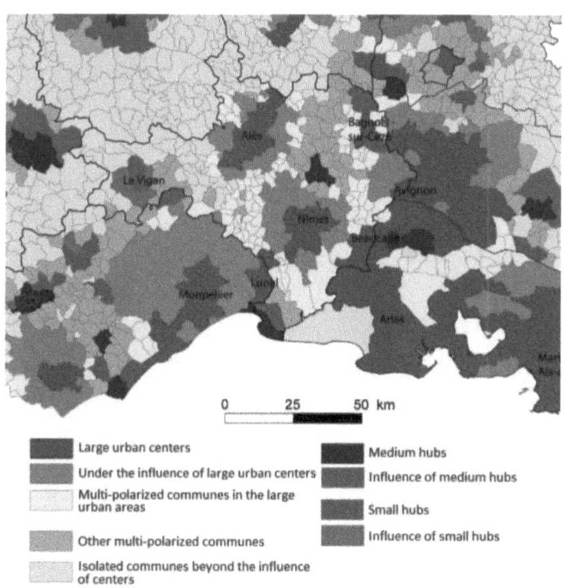

Figure 7.2. *Zoning of urban areas in Gard department and neighboring departments (source: INSEE, Zonage en aires urbaines, 2010). For a color version of this figure, see www.iste.co.uk/lutoff/hydrometeo1.zip*

This urban concentration was accompanied by a partial redistribution of the population toward the south east and the east of the department, two areas that are integrated in the multimodal corridor linking the Rhône valley with the Languedoc arc and Spain (Figure 7.3). The development of industry in Rhône valley allowed many people in the east of the department to find work, notably in the nuclear industry. Similarly, the Languedoc coast benefitted from the development of new tourist resorts. In the context of the Racine mission, six resorts were built in the departments of Pyrénées-Orientales, Aude, Hérault and Gard. The first stones from Port-Camargue in the commune of Grau-du-Roi were laid at the end of the 1960s.

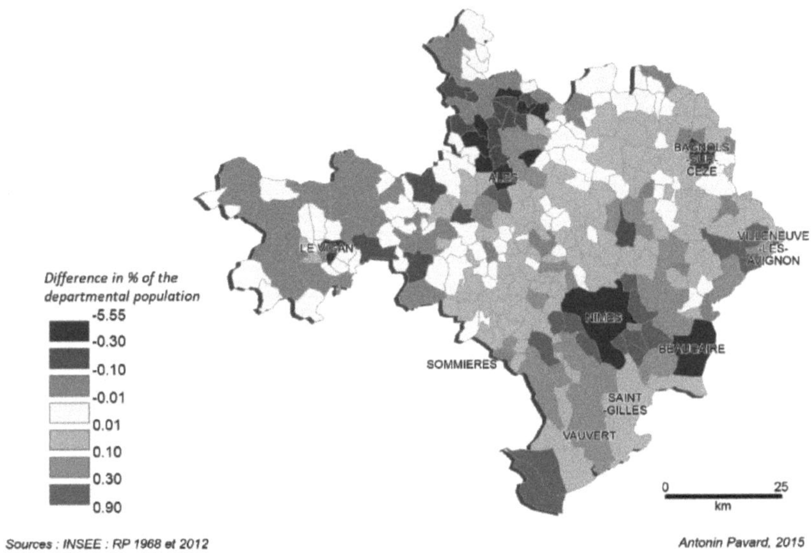

Figure 7.3. *Evolution of the relative contributions of communal populations in the total population of Gard between 1968 and 2012. For a color version of this figure, see www.iste.co.uk/lutoff/hydrometeo1.zip*

On the other side, the region of the Gard Cévennes, located to the west and north west of the department, experienced a notable emigration movement during the beginning of the observed period, due to the rural exodus of the 1960s, and even in the 1970s from the mountains in the south of France, but especially due to the decline of mining activities [CAB 16]. The departure of working-age populations generated a decrease in the number of inhabitants and an ageing of the population which continued into the 1990s. Since then, the communes in this area have shown a more balanced situation, with a slight population growth partially linked to the settlement of neo-country and retired people from other regions or countries, but especially linked with the increase, in recent years, of the number of workers in the agglomerations of Alès, Nîmes and Montpellier [AUZ 01, LEF 09].

The consequence of this movement is an increase in the number of people exposed to the risk of flooding. According to the Gard (Noé) department flood risk monitoring center, nearly 259,000 people live in flood-prone areas, equivalent to around one-third of the population, among which 74,690 live in sectors where hazard levels are high or very high. A comparison between 2007 and 2013 shows the continued increase in exposed populations, in absolute values, but a slight decrease in the percentage (+10,860 inhabitants and –1 point between 2007 and 2013). We can see, reasonably well, the effects of the implementation of flood risk prevention plans (PPRi) during the 2000s and the early 2010s. These did not, however, restrict construction in sectors with moderate hazard levels or the renewal of urban centers.

Analyzing the reasons behind the increase in populations living in flood-prone areas requires the identification of the economic and social factors that pushed the migratory dynamics within the department, and then links them to the evolution of the population exposed to the risk of flooding. However, few articles offer an infradepartmental analysis, notably at the commune level, of the observed migratory dynamics, especially for the beginning of the period retained (1968–2012). Atlases remain the main source of spatialized information [FER 79, DUG 74] but they do not always allow for the establishment of a precise link between the identified factors and the dynamics seen on the maps. However, although monographs are common, they remain difficult to compare and do not always allow the definition of the spatial influence of observed phenomena.

Faced with this double difficulty, we used modeling to estimate the spatial influence of identified factors and their respective weighting in the dynamics of the communes studied. The chosen approach allowed the identification of more localized factors with specific dynamics influencing the evolution of the population also exposed to the risk of flooding.

7.3. The contribution of modeling to the analysis of migratory dynamics in the Gard department[2]

7.3.1. *The role of distance to urban centers*

The analyses highlighted that the main explanatory factor for the migratory rates seen between 1968 and 2010 within the department was the distance to main urban centers. The explanatory power, around 25% at the beginning of the period, nevertheless decreases with time, revealing the growing complexity of residential migrations. To model the dynamics observed within urban areas, we prioritized a

2 The method used is presented in Appendix 5.

slightly revised differential urbanization model. The model used distinguishes three concentric rings that correspond approximatively to the urban center, the suburban peripheries and the nearby periurban areas. It also includes four phases during which the three types of areas identified experience more or less favorable dynamics (Figure 7.4). The identified phases are called urbanization (relative accumulation of populations in the urban centers), suburbanization (accumulation and spreading of a part of the population toward the margins of the agglomerations and the nearby ring), periurbanization (accumulation and spreading of a part of the population toward the villages and towns around urban poles (suburban peripheries) and reurbanization (accumulation and reinvestment of a fraction of the population in the central parts of urban sites).

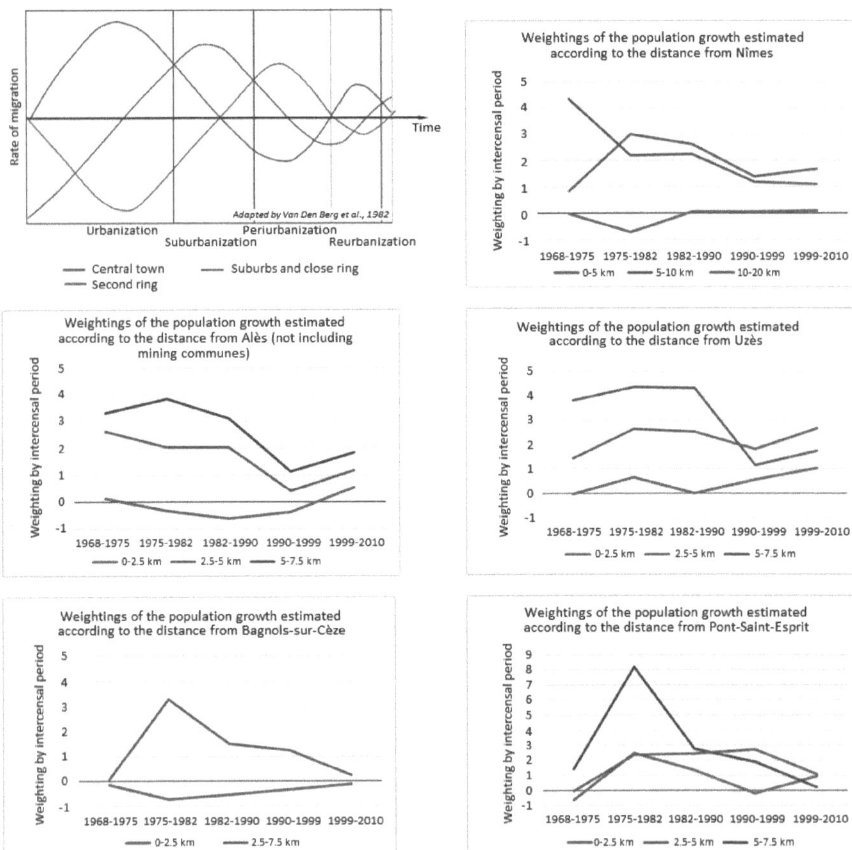

Figure 7.4. *Theoretical model of differential urbanization and its application to the different towns of Gard. For a color version of this figure, see www.iste.co.uk/lutoff/hydrometeo1.zip*

The model was applied to Nîmes, Alès, Bagnols-sur-Cèze, Le-Vigan, Pont-Saint-Esprit and Beaucaire, whose sphere of influence also covers a part of, Bouches-du-Rhône. Lunel and Avignon, at the borders of the department and whose sphere of influence extends over several of Gard's communes, were also retained. Four identical rings (0–2.5 km, 2.5–5 km, 5–7.5 km, >7.5 km) were defined based on statistical analyses, except for Nîmes and Avignon in which the distances covered were greater (0–5 km, 5–10 km, 10–20 km, >20 km) due to a considerably higher demographic weighting.

We can see that the urban centers of Gard's studied municipalities followed, to varying degrees, the different phases identified in the differential urbanization model (Figure 7.4). The different graphs show that during the 1950s–1970s, the urban centers of Gard experienced different magnitudes of population growth. From the 1970s, the aspiration to a single family house was supported by, among others things, an increase in household income, a policy facilitating access to individual property, the motorization of households and the development of a road network, the search for countryside amenities and the positive attitude of peripheral commune mayors (allotment of land, relatively low land costs). For the same period, central communes attracted fewer and even lost populations. Then, the suburban and closest periurban communities became more urbanized and the prices of land and housing increased. Although these migratory movements initially played a role in rejuvenating and increasing the population, the suburban and next closest ring of communes were in turn, over time and with their weak population replenishment, caught up by the aging of their residents. This often gave rise to much smaller populations. The positive dynamics moved further, whereas the price of land was cheaper. In recent years, this sprawling trend was, however, partially counterbalanced. Most urban centers began to show population growth again or, failing that, an inversion of their growth rates. This reurbanization process finds its roots in new policies that aimed to limit urban spreading and to increase the density in towns. It can also be explained by the aging of the population, the reduction of the average household size and, in some cases, phenomena such as the gentrification or, on the contrary, the concentration of vulnerable populations [RER 06].

Finally, the distance to urban centers was combined with the average age of residents, which also played a role in the unequal attractiveness of communes, especially from the 1990s. Since then, two types of commune have become attractive: communes that have benefitted from the establishment of young households, often with children, but also those, often closer to urban centers, in which the average age is relatively high but in which a population renewal is also under way.

7.3.2. *The role of altitude and the distance to Rhône and the sea*

The results of the simulations carried out based on the two factors identified show the important overestimation of populations in communes of Cévennes and their underestimation in the south-eastern plains. We therefore integrated altitude, strongly correlated with the distance to Rhône and the sea, into the model. The choice of this parameter (the altitude) allowed us to restore both the phenomenon of coastal sprawling and attraction to the Rhône corridor and the Languedoc plain in the south of Costières de Nîmes, which concentrates many economic activities, the major transport routes and the Nîmes, Avignon and Montpellier agglomerations. Altitude also allows us to partially take into account migratory dynamics in the central area between Nîmes, Bagnols-sur-Cèze, Alès and Montpellier, which correspond crudely to the Garrigues region in which more favorable dynamics are seen in the plains and in the valleys.

Despite an improvement in the results, they remain unsatisfactory in the north west of the department (to the north of Alès and around Le Vigan), in the Bagnols-sur-Cèze region in the north east and in a whole set of communes located between Nîmes, Alès and Montpellier. Faced with these observations, we once again reviewed the books, articles and atlases to understand the discrepancies between the results observed and those suggested by the model.

7.3.3. *The role of economic activities*

This analysis allowed us to take into consideration the central role of economic activities in the department's migratory dynamics. Several specific cases must be differentiated.

7.3.3.1. *The areas in difficulty*

By examining the history of the communes that were the least well appraised in the model, we were able to see that their economy was often marked by a strong specialization (monoactivity or monoenterprise). Since then, the closure of factories or the restructuring of businesses have pushed populations to move. This is particularly visible in areas marked by a decline in the exploitation of mines, of sericulture and, to a lesser extent, in the confectionary sector.

Mining communities were classified based on the roles they filled: presence of extraction wells (Grand-Combe), of activities linked to the transformation of ores and coal (Cendras and its Blaquière lime factory) or of housing for miners and their families (Branoux-Les-Taillades or Chamborigaud near La-Grand-Combe for example). Three mining areas were thus identified (Figure 7.5). The first, and most important, is composed of the La-Grand-Combe basin and the Cèze valley where many derivative activities were developed. The second groups together the communes around Saint-Laurent-le-Mineur, where the Malines mine is located. The last area, to the west of the department, is close to several mines located on the borders of Gard and the Aveyron (Nants, Saint-Jean du Bruel, Roque-sainte-Marguerite, Revens, etc.). For the communes in question, except for urban centers, we decided to keep the economic factor as the main explanatory variable for the observed dynamics over a large part of the period studied. The same principle was applied to communes marked by the last upheavals of the silk industry. The last mills stopped their activities in the 1950s and 1960s (Barjac in 1957, Saint-Ambroix in 1958, Ganges in 1962 and Saint-Jean-du-Gard in 1965), whereas the end of the 1960s were marked by the restructuring of the silk stocking industry, resulting in the closure of factories in Bez in 1967, or Arre in 1970, and the concentration of final activities in Le Vigan and Sumène. Finally, the temporary decrease in the average annual variation rate observed in the communes of Moussac, Brignon, Sauzet and Boucoiran-et-Nozières is linked to the closure of an important confectionary factory and the concentration of its activities at the Uzès site when the Ricqlès-Zan Company was bought by the Haribo Company in the late 1980s.

7.3.3.2. *The dynamic areas*

The north east of the department constitutes the second sector marked by notable discrepancies between the population estimated in the model and the observed population. The role of nuclear facilities is the main explanatory factor for an observed population greater than the simulated results. Two nuclear sites, about 25 km from each other, constitute the main employment sites for the communes in the north east of the department. They are Marcoule (4,000 paid employees in 2010), located in Chusclan and Codolet near Bagnols-sur-Cèze, and Tricastin, in the Pierrelatte commune in the Drôme department (4,200 paid employees in 2010). Communes in the impact zone of INSEE (Figure 7.6) underwent a demographic evolution strongly directed by the construction of the nuclear sites of Marcoule (1954–1960) and Tricastin (1974–1981), and then by the closure of the Marcoule reactors in 1980, 1984 and 2009, and the closure of the plutonium extraction factory in 1997. In Pierrelatte, the construction of the George Besse II factory for the fabrication of enriched uranium (2006–2010) took over from the George Besse 1 factory, which ceased activities in 2012.

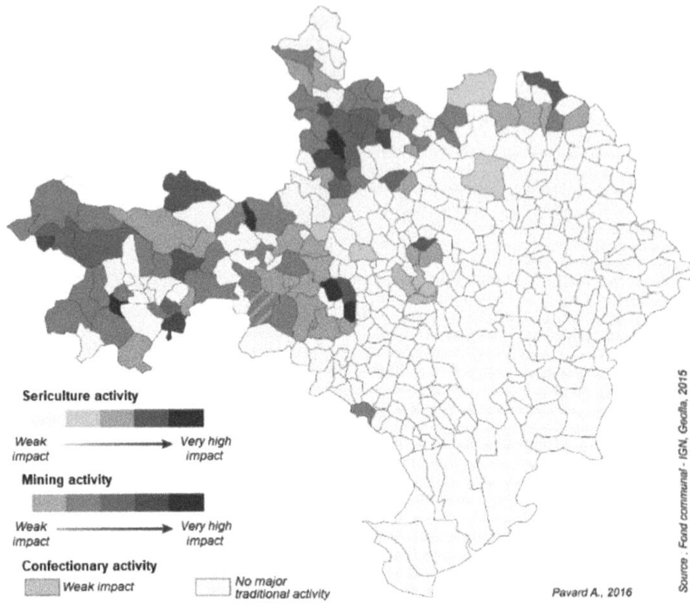

Figure 7.5. *Impact of the decline of traditional activities on the attractiveness of Gard communes. For a color version of this figure, see www.iste.co.uk/lutoff/hydrometeo1.zip*

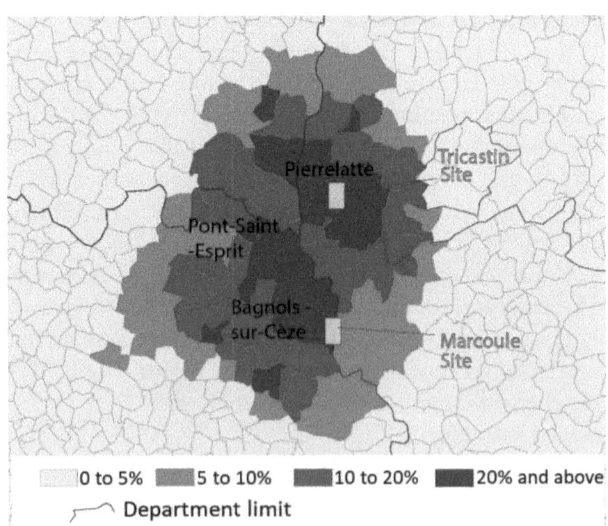

Figure 7.6. *Proportion of commune employment affected by the activities of the Marcoule and Tricastin sites (source: 2010 population census). Adapted from [BON 13]. For a color version of this figure, see www.iste.co.uk/lutoff/hydrometeo1.zip*

7.3.4. *The role of the road network*

The role of the road network in the attractiveness of communes in the department was initially proposed as a determining factor, but the correlation analysis between the distance to a strategic transport axis and the more or less strong attraction of communes was inconclusive. However, it was quickly seen that communes near Nîmes and along the A9 and the D999 or located in a triangle between Alès, Nîmes and Montpellier – benefitting from their proximity to the N106 (Nîmes-Alès), the D6110 (Alès-Montpellier) and the A9 (Nîmes-Montpellier) – experienced a rapid population growth. From this observation, we carried out a typology of the positive impacts of the road network (Figure 7.7), which allowed us to correct some of the discrepancies seen in this area.

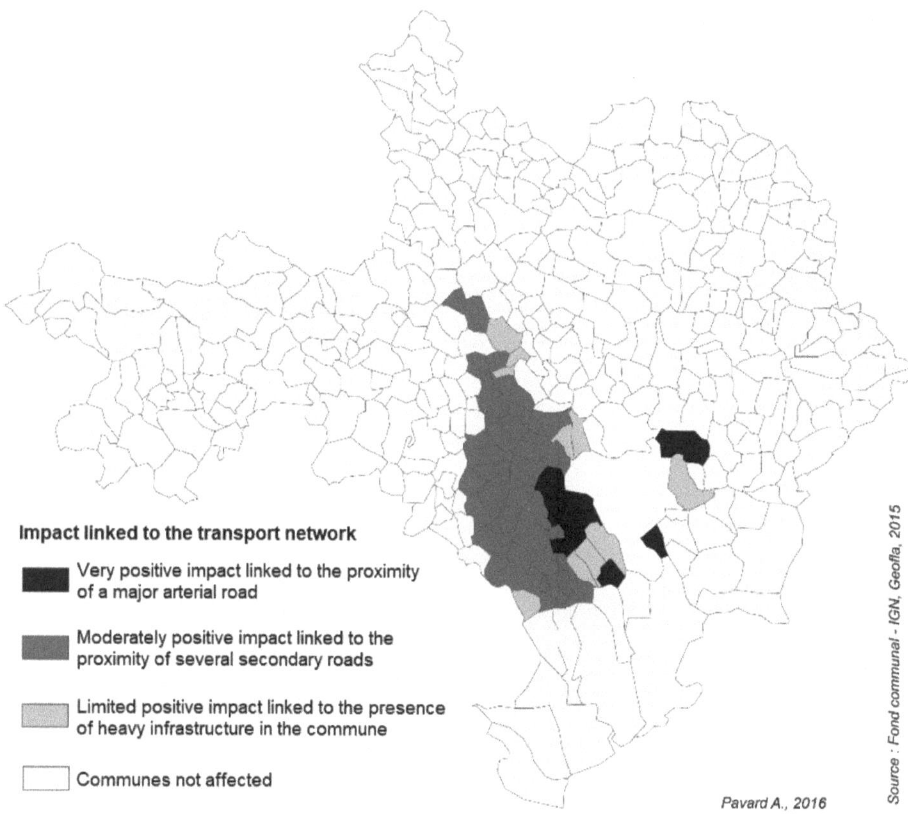

Figure 7.7. *Impacts of the transport network on the attractiveness of Gard's communes. For a color version of this figure, see www.iste.co.uk/lutoff/hydrometeo1.zip*

The articulation of the five factors presented above allowed us to obtain relatively good simulated results. Figure 7.8 is a visualization of the average error per year. A large proportion of the errors still present in the model are due to the two first intercensal periods (1968–1975 and 1975–1982). The following periods show fewer new errors while the last observed period between 1999 and 2010 is modeled best. The cartographic representation of the discrepancies between the observed population and the theoretical population shows that the larger errors are scattered over the whole department. At this stage, it seemed pointless to us to pursue modeling due to the risk of falling into the trap of a monograph (Figure 7.9).

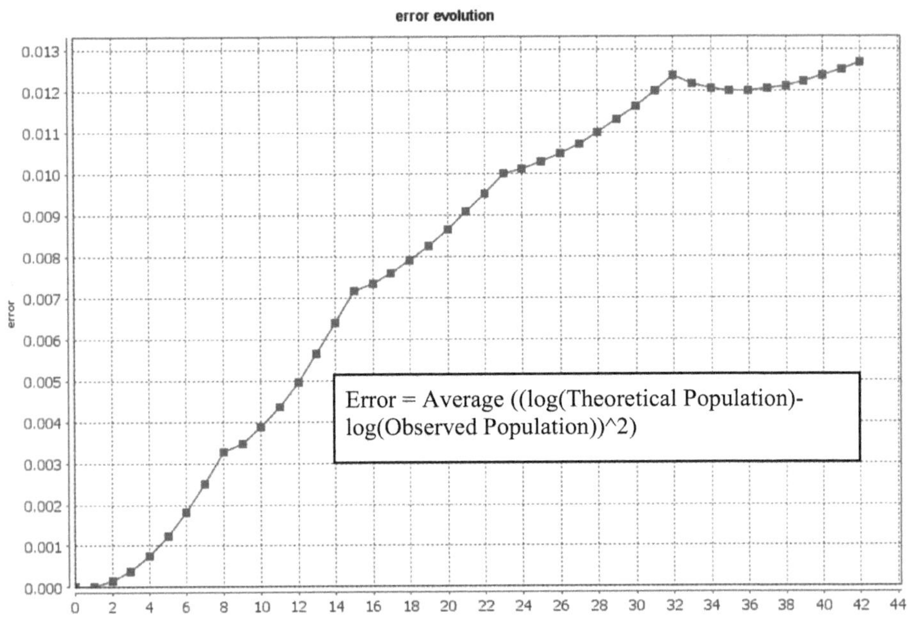

Figure 7.8. *Evolution of the errors accumulated by the model over time. For a color version of this figure, see www.iste.co.uk/lutoff/hydrometeo1.zip*

The model used therefore showed the main factors that pushed the migratory dynamics in the department. A similar analysis was carried out for natural dynamics. It showed that the latter are tightly linked to the average age of the resident population, the distance to urban centers, and the distance to the sea and to the Rhône which is tightly correlated with altitude. Different commune profiles were identified and we can now observe the evolution of the populations exposed to a flood risk since 1968 for each profile.

Figure 7.9. *2010 results of the simulation of population dynamics induced by the migratory balance. For a color version of this figure, see www.iste.co.uk/lutoff/hydrometeo1.zip*

7.4. Analysis of the evolution of the population exposed to the risk of flooding between 1968 and 2013

We used a sample of 33 communes for which we had estimates for the population living in flood-prone areas since 1968 (Figure 7.10). These data, obtained from Gard's flood risk monitoring center, were produced by INSEE based on data gridded for 1968, 1990 and 2006[3]. For 2007 and 2013, the monitoring center used estimates made based on MAJIC data from the "Direction Générale des Finances Publiques" (DGFiP – Directorate general for public finances), used to calculate land tax and to send tax notices. As shown by the Noé monitoring center, changes in the methods used have led to figures that are sometimes very different when comparing 2006 and 2007, especially for communes marked by a notable development of residential areas. It is therefore not possible to compare the number and proportion

3 200 m squares

of the population in flood-prone areas at the beginning and end of the period since the estimation method is different. However, we considered it possible to compare the evolution between two dates using the same estimation method.

The analyses initially focused on the evolution of annual populations in flood-prone areas compared to those in non-flooding areas. Next, we looked at how the exposed populations evolved between 1968 and 1990, 1990 and 2007, then 2007 and 2013 in line with the implementation of regulations restricting growth in risk areas.

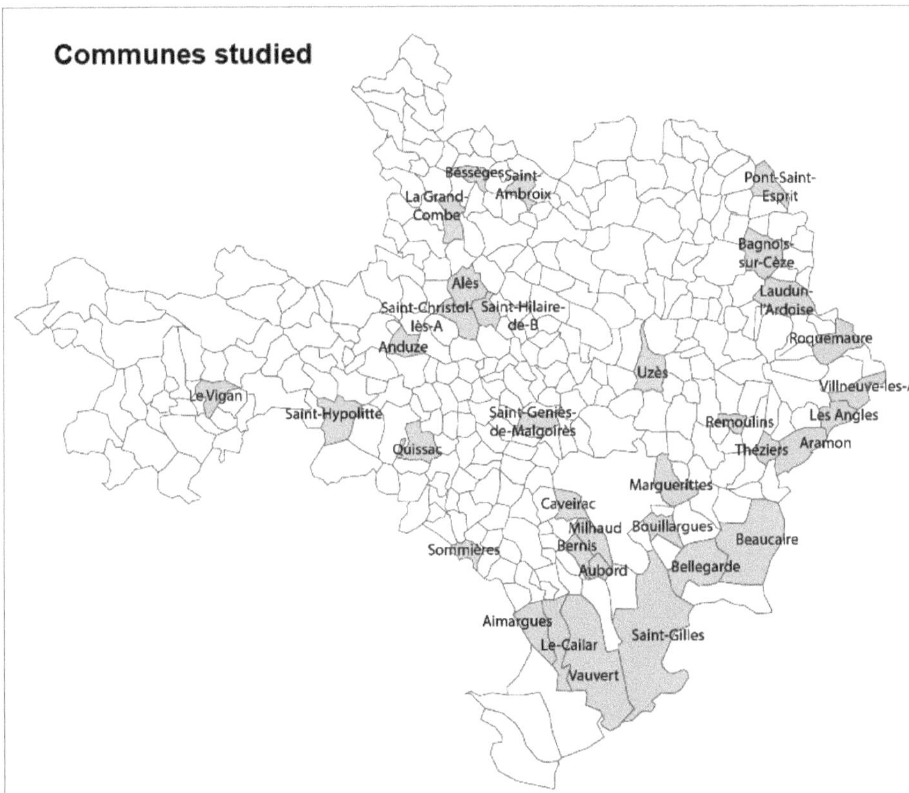

Figure 7.10. *Thirty-three Gard communes observed*

Among the 33 communes observed, the evaluation of the parts of the population living in flood-prone areas in 1968 showed great diversity. At this time, rates varied between 1% in Uzès and Bouillargues and 100% in Aimargues (at the confluence of Vistre and Vidourle). The highest values were mostly seen in the south east, along a route connecting Avignon to Montpellier and integrating communes with a small number of inhabitants in 1968. Only Sommières in the south west and the communes of Roquemaure and Les-Angles toward the north east of this route had more than 3,000 inhabitants.

Although the commune surface area covered by flood-prone areas could explain, in some cases, the proportion of the population exposed, the correlation between these two factors is relatively weak (48%). Communes such as La Grand-Combe and Le Vigan in the Cévennes or Marguerittes and Caveirac near Nîmes have a notable proportion of their population in flood-prone areas, whereas the latter is spatially smaller. Conversely, Beaucaire in the south east, Pont-Saint-Esprit in the north east or Saint-Christol-lès-Alès in the north west are some of the communes with a small proportion of their populations exposed, although flood-prone surfaces cover more than 40% of the commune.

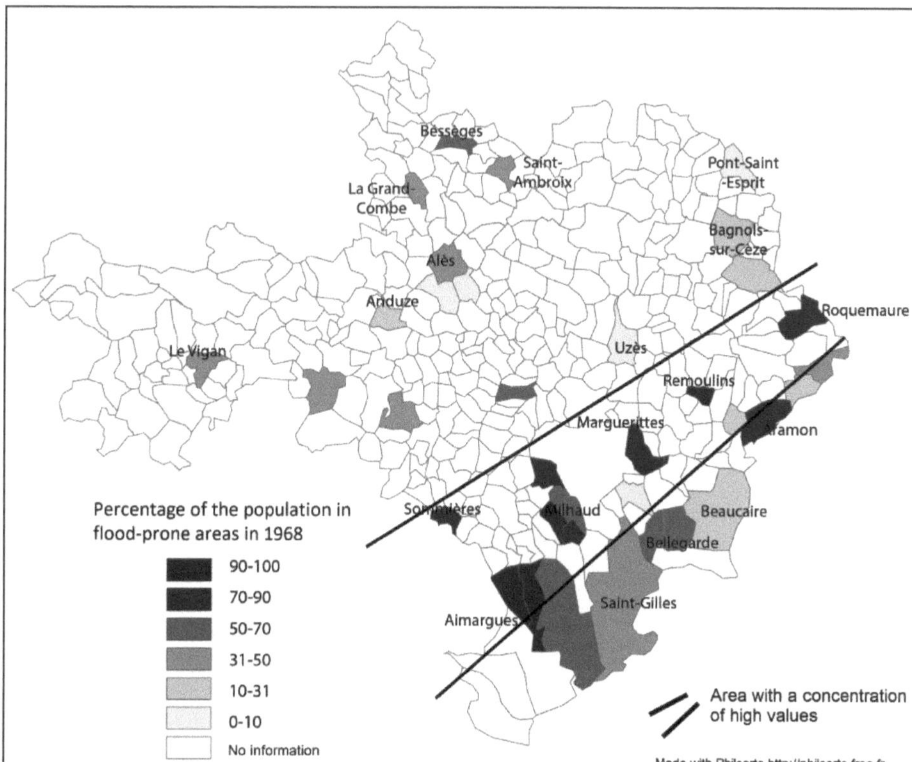

Figure 7.11. *Proportion of the population living in flood-prone areas in 1968. For a color version of this figure, see www.iste.co.uk/lutoff/hydrometeo1.zip*

7.4.1. *Varied trajectories in matters of quantitative evolution*

The 1968–1990 period is the one in which the population in flood-prone areas increased the most. The average annual increase in exposed areas, however, remains slightly less than that of the total population in the 33 communes (5.3% per year against 5.4%). Three types of communes can be identified (Figure 7.12):

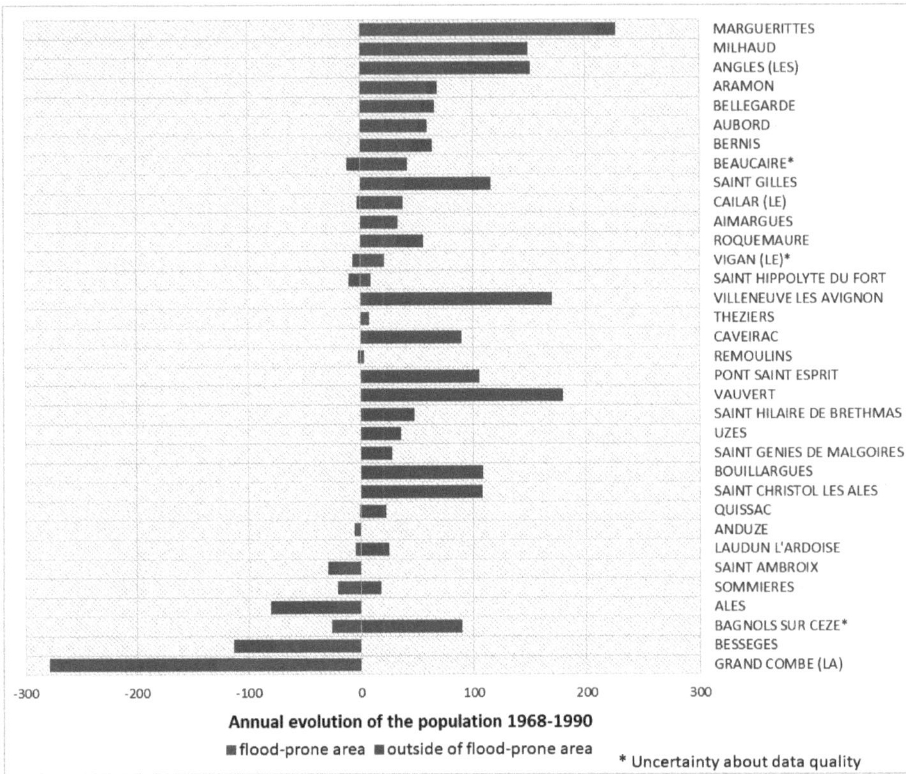

Figure 7.12. *Annual evolution of the population in flood-prone areas between 1968 and 1990. For a color version of this figure, see www.iste.co.uk/lutoff/hydrometeo1.zip*

– those with large rises in populations in flood-prone areas. These are mainly communes that benefit from the urban sprawl of the populations of Nîmes (Marguerites, Milhauds and Aubord), Avignon (Aramon) or located in Petite Camargue between the coast and Nîmes (Bellegarde, Beaucaire and Aimargues). The communes of Les-Angles and Saint-Gilles and, to a certain extent, Bernis can be included in this set but the increase in exposed populations remains moderate compared with the communal evolution;

– those where there is a decrease in the population in flood-prone areas against a background of a decreasing communal population. This set is exclusively constituted of Cévennes communes marked by the difficulties in the mining and textile industry (Alès, Saint-Anduze, Saint-Ambroix and most of all, Bessèges and Grand-Combe);

– finally, a certain number of communes with growth that is sometimes notable show a near-zero evolution of exposed populations. Although this is sometimes facilitated (Uzès, Bouillargues) by a small surface area for flooding, this argument cannot be applied to Villeneuve-lès-Avignon, Saint-Christol-lès-Alès, Pont-Saint-Esprit, Saint-Geniez-de-Malgoirès or Vauvert in the south of the department where the flood-prone surface area covers more than two-thirds of the commune.

The 1990–2006 period was marked by a sharp decrease in the average annual growth rates in flood-prone areas (Figure 7.13). Nevertheless, it remains at 1.1% per year, slightly higher than the evolution of the total population (1% per year).

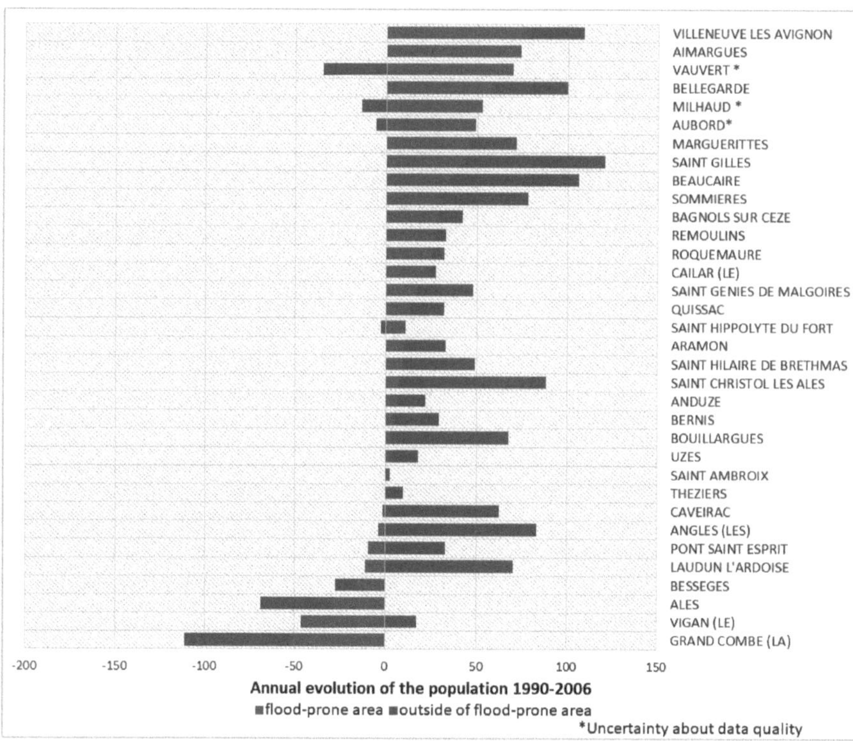

Figure 7.13. *Annual evolution of the population in flood-prone areas between 1990 and 2006. For a color version of this figure, see www.iste.co.uk/lutoff/hydrometeo1.zip*

The most important increase in exposed populations is always seen in the south east where populations are increasingly structured around Montpellier-Lunel, Nîmes and Avignon, to which the secondary center of Beaucaire can also be added. Similarly to the preceding period, the loosening of the Nîmes population favors a sharp increase in the exposed population in Milhaud, Aubord and Marguerites, whereas Bernis, Bouillargues and Caveirac continue to limit the evolution of populations in flood-prone areas. Important changes affect the communes under the influence of Avignon. Villeneuve-lès-Avignon, which was able to control the evolution of its population in flood-prone areas between 1968 and 1990, then experienced one of the most notable increases among the communes studied. Conversely, Aramon, Théziers and Les-Angles have increasingly controlled settlement in flood-prone areas against a background of a strong population growth.

The same is seen in the communes of Saint-Christol-lès-Alès, Saint-Hilaire-de-Brethmas and Anduze near Alès, much like in Sommières, Quissac and Saint-Geniès-de-Malgoirès, which have undergone a population growth supported by their intermediary position between Alès, Nîmes and Montpellier and by the improvement of the road network. In the north-east, Pont-Saint-Esprit and Laudun-l'Ardoise, bordered by the Rhône, have a policy that remains rigorous in terms of restricting settlement in flood-prone areas while the majority of communes in Cévennes continue to decrease in size, favoring a decrease in the exposed population.

The most recent period (2007–2013) shows a notable decrease in settlement in flood-prone areas, demonstrating the effectiveness of the flood risk prevention plans that affect a large proportion of the communes studied (whether recommended or approved) and perhaps a greater awareness of the risk following several dramatic floods (2002, 2003 and 2005). The average annual growth rate in the flood-prone area is now markedly less (0.7%) than in the population of the 33 communes where a small increase of 1.2% per year is seen.

Beyond the commune of Aimargues, which is fully in a flood-prone area, only Milhaud and Aubord, near Nîmes and Bagnols-sur-Cèze in the north east, show an expansion that is mostly focalized in flood-prone areas (Figure 7.14). The majority of population growth is seen in non-flooding areas, including in La-Petite-Camargue, which covers the communes located in the south of Nîmes.

These different elements show a wide variety of trajectories. While certain communes such as Pont-Saint-Esprit, Laudun-l'Ardoise or Uzès demonstrate a willingness to continue limiting the population growth in exposed areas, even against a background of a sometimes very high population growth, many communes with high growth rates alternate between severity and flexibility. This affects most of the communes located within the sphere of influence of Nîmes, Avignon or Alès, that have to manage the large influx of new inhabitants, as well as certain communes of La-Petite Camargue. The decrease in populations in the flood-prone areas of Cévennes has occurred in a context of a marked population decrease, although the situation appears more favorable between 2007 and 2013 than before.

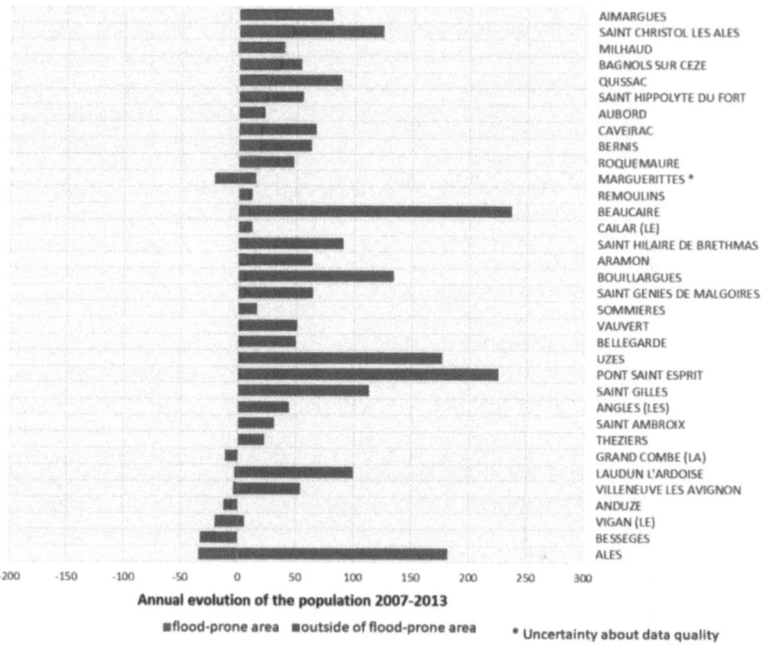

Figure 7.14. *Annual evolution of the population in flood-prone areas between 2007 and 2013. For a color version of this figure, see www.iste.co.uk/lutoff/hydrometeo1.zip*

7.4.2. *A recent trend but generalized to the decrease in the proportion of the population exposed*

A large proportion of these crude evolutions are influenced by the size of the communes. This is why this analysis is complemented by a study of the evolution of the proportion of the population in flood-prone areas. It is useless at this point to review cases of communes in Cévennes that are marked by a specific history. We propose to place a greater focus on the communes located within the spheres of influence of the main towns and those located along the Rhône in order to link the variations observed with the establishment of a regulation aiming to restrict construction in flood-prone areas.

The data confirm the diversity of situations in communes close to Nîmes and Alès who have recent (2014 and 2010) flood risk prevention plans (PPRi). Around Nîmes, the communes of Margueritte and Milhaud are two of the rare communes in which the proportion of exposed populations increased between 2007 and 2013 while, in Aubord in 2013, nearly the entire population was in flood-prone areas after a large increase between 1990 and 2006. In contrast, the proportion either remained stable or decreased in Bernis, which also has an extensive flood-prone areas, Caveirac which has implemented the R111-3[4] regulation since 1996 and to a certain extent, in Bouillargues (Figure 7.15), attesting to an increased awareness of flood risks. This diversity of situations is also seen when Saint-Christol-lès-Alès and Saint-Hilaire-de-Brethmas, which adjoin Alès, are compared. These two communes have a similar number of inhabitants in 1968 and have comparable flood-prone surface areas. Although the latter was able to contain the proportion of the population exposed, the same was not the case in the former that became increasingly urbanized over the 45 years observed.

Conversely, we observed a certain homogeneity in the communes near Avignon: they all showed a relative decrease of populations in flood-prone areas, from 1968 to 1990 in Roquemaure, between 1990 and 2006 in Les-Angles and Aramon and between 2007 and 2013 in Villeneuve-lès-Avignon. More broadly, all the communes sampled along the Rhône showed an early decrease in the proportion of the population exposed. As of 1982, the majority was included in a submersible surface plan[5]; Roquemaure and Laudun-l'Ardoise were also the first communes in Gard to implement a PPRi in 2000. The rapid establishment of flood regulation enabling the control of urbanization in flood-prone areas therefore turned out to be essential. Two other communes implemented the R111-3 regulation before the establishment of their PPRi: Le-Vigan in 1998 and Le-Cailar located at the confluence of Rhôny and Vistre in 1994, then in 1996. Here again, we observed a significant decrease in the population exposed from the beginning of the 1990s, even though the proportion remained at 92% in Cailar in 2013.

Finally, it is interesting to note that agricultural Camargue (Beaucaire, Bellegarde and Saint-Gilles) has also controlled the evolution of exposed populations since 1990. Since the regulation came in quite late (2012), it is undoubtedly due to the effects of the 1993 and 1994 floods that strongly affected this area.

---

4 The R111-3 stipulates that construction on lands exposed to a risk, such as flooding, erosion, collapse, landslides and avalanches, can, if authorized, be contingent upon special conditions.

5 Angles, Villeneuve-lès-Avignon, Laudun-l'ardoise, Roquemaure and Pont-Saint-Esprit, the latter having also been affected by the PSS of Ardèche in 1959.

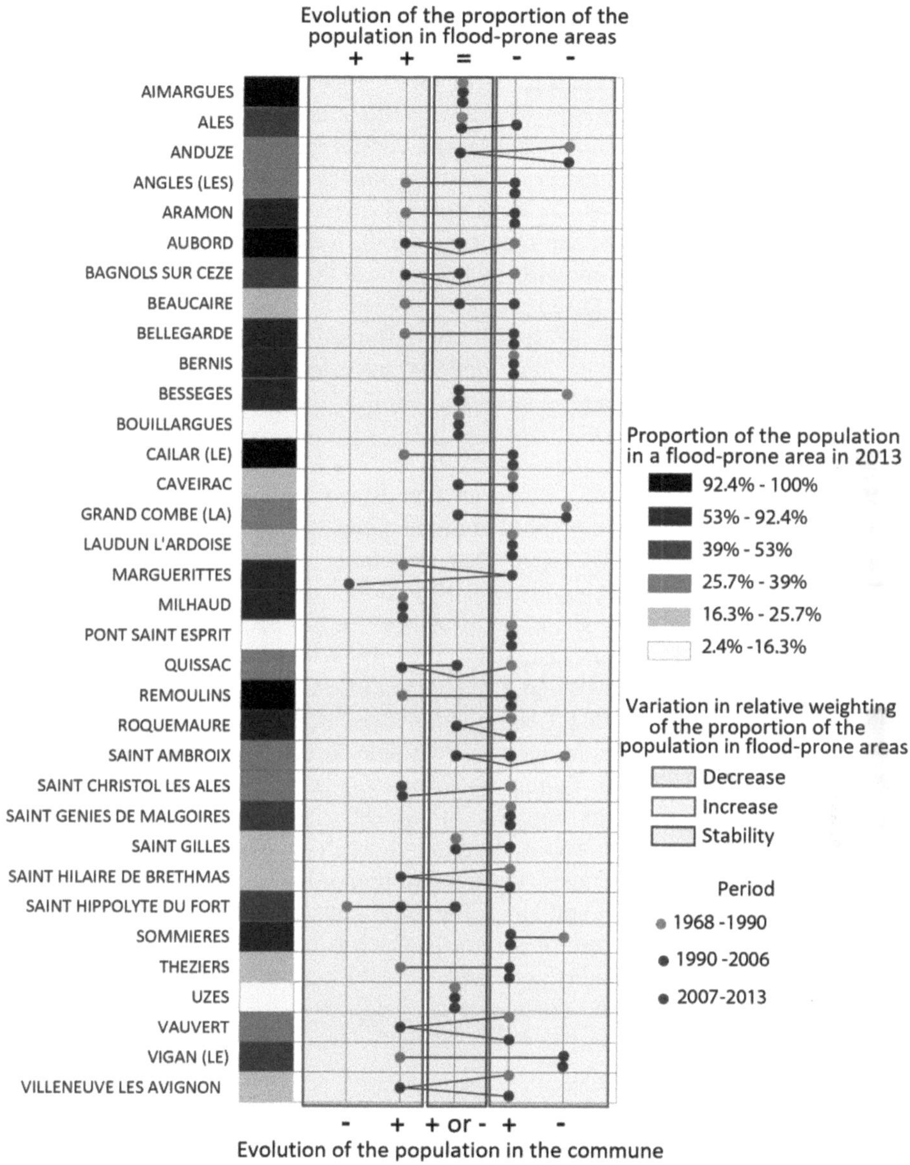

Figure 7.15. *Evolution of the proportion of the commune population in flood-prone areas between 1968 and 2013. For a color version of this figure, see www.iste.co.uk/lutoff/hydrometeo1.zip*

7.5. Conclusion

The population dynamics observed in Gard since 1968 have essentially been pushed by population migrations. These migrations arise from trends seen at the national level such as the periurbanization and coastal sprawl of populations, and the attractiveness of regions located in the south of France. They can also be explained by more local factors such as the development of nuclear facilities in the Rhône valley or the improvement of the road networks in certain geographical area. The construction of a computer-simulated model allowed the precise spatialization of the different factors identified. Discrepancies between the simulated results and the reality served as a guide in the search for information that was often scattered. Several commune profiles were thus defined and served as a framework for the analysis of the quantitative and relative evolution of the population exposed to the risk of flooding.

From a general point of view, it appears that if the population in flood-prone areas has not stopped increasing, the proportion of the population exposed tends to decrease, especially after 2007, highlighting a more favorable situation. To achieve this, a certain number of communes used regulatory tools, sometimes pushed by national specific regulations and well before the establishment of PPRi. This was the case for the communes bordering the Rhône which, from 1982, were subject to a submersible surface plan and to which Le-Vigan, Caveirac and Le-Cailar, where the R111-3 was used in the 1990s, can be added. Certain communes such as Bernis, Saint-Hilaire-de-Brethmas or Saint-Geniez-lès-Malgoirès were able to control the proportion of the population exposed without falling back on regulations, despite the extensive flood-prone surface area. In contrast, it appears that the establishment of a regulation was necessary for certain communes such as Marguerittes, Milhaud or Saint-Christol-lès-Alès where an increase in the proportion of the population exposed was still noted between 2007 and 2013. For these communes, the delay between the date of prescription and the date of approval (which is recent) leads to the suggestion that negotiations on the regulatory zone were the subject of debate.

7.6. Bibliography

[AUZ 01] AUZEBY F., LE GOUHINEC T-D-T., "Migrations de retraités en Languedoc-Roussillon", *Repères pour l'économie du Languedoc-Roussillon*, no. 15, INSEE, 2001.

[BON 13] BONNET M., DUPRE A., "Les sites nucléaires du Tricastin et de Marcoule : un impact important sur plus de 200 communes", *La lettre analyses,* no. 188, INSEE, 2013.

[BUF 16] BUFFARD P., "Panorama du Gard – Le 3e département de la région le plus peuplé", *Analyses Occitanie*, no. 8, INSEE, 2016.

[CAB 16] CABANEL P., *Histoire des Cévennes*, Presses Universitaires de France, Paris, 2016.

[DUG 74] DUGRAND R., FERRAS R., JOUTARD P., *Bas Languedoc, Causses, Cévennes*, Larousse, Paris, 1974.

[FER 79] FERRAS R., PICHERAL H., VIELZEUF B., *Atlas et Géographie du Languedoc et du Roussillon*, Collection Portrait de la France moderne, Flammarion, Paris, 1979.

[LEF 09] LEFEBVRE M., FOYER-BENOS B., "Le Parc national des Cévennes: Regain démographique pour un territoire ouvert sur l'économie environnante", *Repères Synthèses pour l'économie du Languedoc-Roussillon*, no. 8, INSEE, 2009.

[OBS 15] OBSERVATOIRE DU RISQUE INONDATION DANS LE GARD, "Estimation de la population résidant en zone inondable dans le Gard", *Les publications de l'observatoire du risque*, no. 16, 2015.

[RER 06] RERAT P., "Mutations urbaines, mutations démographiques: contribution à l'explication de la déprise démographique des villes-centres", *Revue d'économie régionale et urbaine*, vol. 5, pp. 725–750, 2006.

[VAN 82] VAN DEN BERG L., DREWETT R., KLAASEN L. *et al.*, *Urban Europe: A Study of Growth and Decline*, Pergamon Press, Oxford, 1982.

8

Exposure to Flash Floods: The Conflict Between Human Mobility and Water Mobility

8.1. Introduction

The heart of the flooding problem is human exposure. A society is exposed to floods when the individuals within it, as well as their goods, are subject to the overflow of a watercourse. By "heart of the problem", we mean that exposure is the component of risk at the interface between hazard and vulnerability, where the social and physical dynamics converge. Exposure must therefore be studied as a coupled human-natural system [LIU 01].

Human exposure to floods is a sociohydrological[1] problem in the sense that it results from a dual conflict between human mobility and water mobility, which usually coevolve pacifically [SIV 12, SIV 15]. When water invades the spaces used for human activities, human activities attempt, in return, to limit this invasion. Two attitudes are then observed and develop over both the short term of the event as well as over the longer term for prevention. We use the term *resistance* to characterize activities that aim to control the mobility of water. In the long term, we speak of

Chapter written by Céline LUTOFF, Jean-Dominique CREUTIN, Isabelle RUIN, Sylvie DUVILLARD, Sandrine ANQUETIN and Marco BORGA.

1 The alteration of economic, political, technological and social processes after a flash flood is typically a sociohydrological problem requiring sophisticated description and modeling [DIB 13, DIB 15].

active protection[2] (i.e. structural protection through the construction of dikes); in the short term of the event, the behaviors for resistance can lead to the establishment of cofferdams for example. The second attitude is that of *acceptance*. This is shown through actions that aim to withdraw activities from areas that are invaded by water, therefore working in terms of human mobility. In the long term, this could be non-structural protection such as the establishment of constraints and constructive recommendations in flood-prone areas for example; at the time of the event the acceptance behavior can be seen in the evacuation of areas and cancelled travel or avoidance of flooded roads.

Human mobility can be broken down into (1) long-term individual mobility driven by the residential choices, whether issued through public or private decisions, and (2) daily mobility regulated by individual schedules involving movement between areas of residence, work, hobbies, business, etc. Daily mobility is influenced by residential mobility, and both evolve according to different temporal characteristics, from the hourly to the annual scale. Water mobility also includes two components: (1) the dynamic of the river in the long term (both in terms of flow and the morphology of the river) governed by the frequency of floods, which are themselves dependent on the large-scale hydroclimatic processes, and (2) the short-term dynamic of the river in overflow, controlled by the event's meteorological situation and the initial hydrological conditions [BLO 95].

Human and water mobility interact and regulate each other. These interactions and reactions cross spatial and temporal scales, making their apprehension complex. This is the task that this chapter aims to tackle. It aims to illustrate these multiscale interactions and reactions between human mobility and water mobility by combining physical and social data [RUI 14]. We place a particular focus on the temporal scales engaged by these interactions, by illustrating our statements across the whole range of spatial scales, from the local to the global level. It should be noted that these spatial and temporal scales are not implicitly correlated: the short term does not exclusively involve the local level, and the long term does not necessarily apply to the national level for example. We will support our analysis using data accumulated over several interdisciplinary projects on flash floods [RUI 07a, CRE 09, CRE 13, TER 15, LUT 16, SHA 17]. The analyses presented here, from these projects, are based on fieldwork in the Gard (around 5,900 km^2 and 700,000 inhabitants) and Var

2 Di Baldassarre *et al.* [DI 15] describe the balance between active protection and passive protection as equivalent to a choice between erecting dikes (technological side) or creating a distance to a watercourse by restricting the development of the population (economic side) and the awareness of risks (social side). On the notions of active or passive protection, we also fall back on the works of Van Dine [VAN 96].

(around 6,000 km^2 and 1,040,000 inhabitants) Mediterranean departments[3] in southeastern France, where flash floods constitute the most critical risk of natural origin [BOR 08, BOR 11]. The Mediterranean region is actually one of the most sensitive areas on the planet in terms of both the concentration of climate change effects [PLA 08] and the strong and growing demographic pressure leading to constant urban spreading[4].

Flash floods occur over short time periods and space scales (catchment areas of 2–2,000 km^2) and condense a high level of energy (flow rates of up to 10–20 $m^3 \cdot s^{-1} \cdot km^{-2}$ for hydrographical basins of up to 100 km^2). They generally affect surface areas limited to a few thousand km^2 with storm rainfall accumulations of more than 200 mm in 24 h (as was the case in the Gard watershed during the events of September 8 and 9 2002 [DEL 05]). Compared with slower floods, this unfortunate conjunction of meteorological and hydrological phenomena highlights the conflict between the previously described mobilities in at least three ways:

– human exposure is dependent on residential choices in the long term but also on mobility choices in the short term of the event since pedestrians and drivers are exposed as much as residents;

– exposure problems lie on different levels of magnitude, from a few minutes for travel decision-making when faced with flooded roads to several decades for institutional response to the flooding phenomena;

– exposure is more easily reducible by acting on human mobility rather than on water mobility, due to the size, violence and number of threatening watercourses.

In order to present the main results of our work, Table 8.1 aims to illustrate the different aspects of interactions between human mobility and water mobility. The two previously described possible options of *resistance* (option a) and *acceptance* (option b) are represented in the short term of the event and the long term of prevention. It is understood that people can pass from one option to the other at any timescale, from minutes to decades. Type (a) interactions aim to preserve human enterprises from hydrological disturbances by referring to an attraction/resistance process with regard to water, whereas type (b) interactions relate to a process of flood repulsion/acceptance and aim to maintain a secure distance from the water.

3 A French "department" is an administrative unit belonging to one of the three levels of government below the national level, between the administrative regions and the commune.
4 On the question of the attractiveness of areas at risk in this region, see the works of Durand [DUR 15a].

	Short hydrological time: the event	*Long hydrological time: inter-event*	
Short-term human mobility	a) Refusal to evacuate, driving on flooded roads, temporary protection	a) Protection of roads	Vehicle Daily mobility
	b) Alert system, evacuation, road closure, travel modification	b) Choice of route itinerary, anticipation of the implementation of alternative itineraries or modes of transport in case of flooding	
Long-term human mobility	a) Repair or reconstruction of damage	a) Protection infrastructures (dikes, dams, etc.), design of waterproof buildings	Accommodation Residential mobility
	b) Sale of houses or demolition, delocalization	b) Spatial planning, building permit requests	

Table 8.1. *The potential interactions between flooding and human mobility over the long and short term*

The interaction between the occurrence of a flood and daily mobility essentially concerns the actions undertaken to accommodate daily activities to the flooding circumstances. The interaction between the occurrence of a flood and residential mobility concerns the individual and collective decisions related to residential choice after a flood event. For each square, two components of each of the options offered to individuals and groups are distinguished: the modification of water mobility (so-called active protection) or the adaptation of their mobility (so-called passive protection).

The chapter follows the same logic as Table 8.1. Our analysis focuses, first, on the interactions during the event (short hydrological time, section 8.2), then on those relevant to the interevent period (long hydrological time, section 8.3), each time taking into consideration the actions of water on human decisions and vice versa. In section 8.4, we illustrate the cross-scale interactions between short and long (diagonal) timescales. Finally, in a final conclusive section 8.5, we put forward a multiscale representation of flood exposure by highlighting the need for research on the intermediate timescale of "normal floods".

8.2. The rapid dynamics of flood–society interactions: the time of the event

Here, we aim to better understand how the occurrence of a flood can disturb and modify the routine activities of individuals by recording the sequence of actions implemented by the individuals at the time of the crisis. During a crisis, the interactions between the individuals and the flood are governed by the rate of the rising water. The usual time for the development of a flood is linked to a power law that depends on the size of the watershed in question [MAR 10]. Given the size of the catchments affected by the flash floods (from 2–2,000 km^2), the usual response time (time between the peak of the rain and the peak of the flood) can vary from a dozen minutes to several hours, but is always less than a day. With a myriad of subcatchments reacting to the organization of convective cells that generally cover several thousands of km^2, a flash flood exposes individuals in an asynchronous and heterogeneous manner over the entire affected area[5]. The challenge in terms of human exposure is therefore to understand how the changeover between routine activities and actions undertaken in response to floods is done, in the specific timeframe of the event according to the place where each individual is. It is on this point of the changeover that this section is dedicated. To do so, we relied on 29 interviews undertaken following the catastrophic flood of June 15 2010, in Var. The sequence of actions reconstituted based on these interviews refer to the most dramatically affected watershed (watershed of the Naturby River, 200 km^2 in Draguigan), where the majority of the 26 deaths from the event occurred[6].

8.2.1. *Accept or resist*

Acceptance covers situations in which the principle component of the action is to remain at a distance from the flood [RUI 07a]. At the individual level, this could translate into actions such as leaving work early, reorganizing a schedule or adapting the car journey to avoid flooded roads. Anticipated evacuations or sheltering by moving to higher levels or placing objects at a height before the arrival of the water are additional examples. Closing floodable roads by redirecting drivers to alternative routes or closing schools and canceling school transport to anticipate the flood are examples of actions taken at a more collective level. At this level, public transport services could also be suspended or modified to anticipate the occurrence of a coming major hydrometeorological event.

In contrast, *resistance* covers situations in which the actions undertaken – although they can be qualified as “protection” – aim to allow the continuation of planned

5 On this notion of asynchronicity and scale, we refer to [CRE 13].
6 Ruin *et al.* give a detailed description of this event for which radar monitoring showed 200 mm in 10 h on the Naturby watershed [RUI 14].

activities, with the notion that the flood can be controlled. Behaviors such as refusing to evacuate or driving on a flooded road implicitly translate the idea that the building or vehicle is resistant enough to deviate the flow of water [RUI 07a]. Similarly, protecting the entryway of a building using sandbags, or a door with a cofferdam returns to the supposition that reinforcing a structure is enough to deviate the water. These resistance methods can be improvised at the needed time or be anticipated. As such, we find increasingly adapted (removable, inflatable, etc.) individual protection measures on the market to prepare for the eventuality of a flood.

Without prejudging the good or bad, prudent or risky character of such behaviors, we consider these protection actions as clear signals of a flood response. We will now analyze how the transition between "routine" actions and those "triggered by a flood" is done, notably one of the first protection actions aiming to leave the flooded space.

8.2.2. *A delicate transition between routine activities and awareness of the event*

In order to illustrate the way in which individuals go from their daily activities to actions undertaken to face a flood, we asked flood victims to describe the sequence of their actions during such events. Based on these descriptions, we referred to what we call a "flood anticipation scale" [LUT 16], grading the passage from habitual daily activities (or planned for the day in question) to activities undertaken due to the current flood. From the data collected, we established that this passage occurs in three stages: the search for "information", "organization" actions and "protection" actions. Of course, the boundary between the two categories of activities ("routine" versus "triggered by a flood") is not as clear in reality, and individuals generally fluctuate between them, alternatively seeking to be informed, to organize themselves, to protect their goods, all while attending to their daily activities. It takes time for each individual to realize and accept that the flood will really affect them, and that it is time to react [RUI 07a, RUI 09]. Through our categorical simplification, we consider that the passage to taking the flood into account is clear when individuals undertake actions for their protection, whether this consists of *resisting* (by attempting to deviate the flow of water) or of *acceptance* by adapting (and heading to a shelter).

The testimony of two of the people questioned who experienced similar flood conditions in the same area during the floods of June 15, 2010, clearly illustrate the two types of protection behavior. The first person, a 23-year old civil servant, is particularly representative of acceptance behavior. At work during the event, in an office on the first floor of a building next to the Naturby River, this person was able to observe the rise in water and see unusual signs such as objects floating in the river. Faced with the worrying situation, he "organized" his response by warning his superior and his colleagues. He then placed objects in a safe place on higher shelves before leaving his office to warn employees in neighboring businesses of the danger, urging them to evacuate immediately. Here, we have an example of a person who linearly modifies his activities on our "anticipation scale": he observes the warning signs that lead him to accept, very early on, the potential danger of the situation, and to then consequentially adapt his behavior by quickly moving away from the river and inviting those in his vicinity to do the same. This sequence of actions is therefore easy to code, as long as the timelines specified in the interviews are correct, which they usually are in comparable situations [GAU 04].

The second witness is a shop owner in his 30s. At the time of the event, he was concentrating hard on the shelving for newly received products. It was only when there was 30 cm of water in his shop, and after being alerted by two people in his surroundings (of which one was the previously mentioned civil servant who said to him: "if you do not evacuate, you will die!") that he was finally convinced of the need to evacuate – him and his two employees – and head to the first floor of the neighboring building. Here we have an example, very different from the first, of a person who, suddenly, accepts the need to adapt to the worrying rise in water, having first adopted a resistant behavior and maintained his routine activities, presuming that his shop would protect them – him, his employees and his goods. The civil servant mentioned above summarized this behavior by saying that "they picked the wrong battle". This sequence is also easy to code in terms of the passage from "routine" activities to activities triggered by the flood.

8.2.3. *Chain reactions in flooding tempos*

The 29 testimonials describing the sequence of actions undertaken during the event also allow the quantification of the speed of human reactions when faced with flooding. Based on these accounts, we can visualize the total time needed for all the people questioned to adopt protective actions. We can also link the timeline of these protection actions to the peak of the flood to which they refer, thus determining a measure of anticipation for the actions. This last point is important in the sense that an allegedly adaptive action might turn out not to be if undertaken too late in terms of the rising water. For example, placing objects at a higher level or parking a car at a high point are behaviors that fall in the "acceptance" category. However, if these actions are undertaken with "feet in the water", they can quickly become dangerous, especially in cases of flash flooding.

Adapting one's own rhythm and actions to the apparent hydrometeorological signs is key to protecting oneself, and very fortunately, this occurs in the majority of cases. This is what is observed in Figure 8.1, which represents how the sample of people questioned, exposed to the same storm that generated flash floods, switched from their daily activities to protection activities. In terms of the contexts seen, some people were at work, others at home or on the road; most of the protective reactions reveal a relatively adapted dynamic with a relatively short anticipation time (around an hour) and a good rhythm of achievement of protective measures (around 2 h to achieve a 100% modified sequence of actions and a phase to return to normal beyond the first hour).

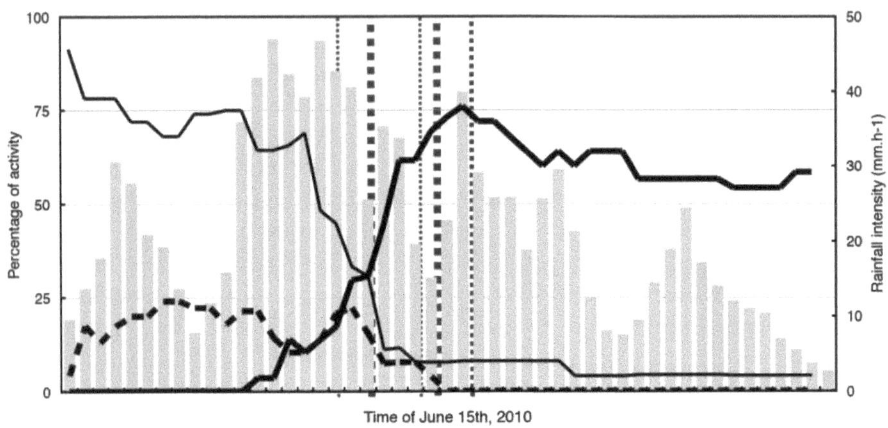

Figure 8.1. *Coevolution of hydrometeorological conditions and sequence of individual actions during the event of June 15, 2010, in Draguigan (France), reconstituted based on postevent interviews with 29 flood victims. The gray bars represent the evolution in average intensity of rain (mm·h^2) over the affected area (watershed of 200 km^2). The horizontal dotted black lines represent the hours of the peak rise in water in the Naturby River (184 km^2 and 196 km^2) and the tributaries of Riaille (6.2 km^2) and Tuilière (4 km^2 and 19 km^2); the thickness of the thin, medium and thick lines reflect the surface areas of the watersheds. The vertical black lines represent the percentage of people classed for each increment of time according to the category of activity under way: (1) information–organization (dotted), (2) protection (thick line) and (3) routine activity (final line)*

In this figure, we see that before switching from normal daily activities to protective actions, individuals tend to search for information in order to understand the situation, to try and evaluate their options and organize their consequent behavior (the first sequence of rain corresponding to the information–organization phase). Based on this sample, the response time between the acquisition of information and the implementation of protection appears to be around an hour.

8.3. The slow dynamics of flood–society interactions: inter-crises

Society–flood interactions are governed by the occurrence rate of floods over time. The average time in France between two catastrophic events is around half a century to a century for an area with a given risk [LAN 13]. The spatial scale of an event – and the risk extension area – depends on the type of flood: usually less than several thousands of km^2 for flash floods. In this section, we look at the inter-crises interactions at two levels – local and national – and always consider these interactions according to the two previously mentioned dimensions: the active and passive types of protection.

For the local level, we use our research based in the Gard department (around 5,900 km^2 and 730,000 inhabitants), regularly affected by flash floods that sometimes cover the entire region [DEL 05]. The area studied is located between the Cévennes mountain range (whose summit stands at 1,500 m) and the Rhône delta (located at sea level), which is mainly drained by the river Le Gardon, a tributary of the Rhône. Ceresetti *et al.* [CER 10, CER 11] and Moliné *et al.* [MOL 12] showed that, based on long-term pluviometry observations, this region is subject to a rainfall pattern that produces hourly and daily accumulations close to world records. High-resolution atmospheric modeling has shown that these accumulations are mainly caused by near-stationary mesoscale convective systems [NUI 08, NUI 09] (see Chapter 3).

At the inter-crisis level, human mobility is apprehended through residential mobility, documented through the tax declarations to which property owners are subject to at the end of construction. The data from these declarations contain different kinds of information about the property in question (surface area and position relative to the official flood-prone area), the building (type of usage and surface area) as well as the proprietors (sociodemographic data). Given the administrative and practical constraints (delays in validating building permits, for the design as well as the construction), this declaration is usually established 3 years after the award of a building permit. This means that the data lags about 3 years behind the decisions of interest to us: those of construction, or not, in flood-prone areas.

At the national level, we examined the slow evolution of water mobility – as for all French regions with a flood risk – through two main sources: (1) the more or less well-organized rain systems covering small watersheds (from 2 to 2,000 km^2 – rural or urban) and characteristic of the Mediterranean regions and (2) the extra-tropical long-lasting or repetitive cyclones that affect the larger French catchment areas (from 10,000 km^2 to 100,000 km^2). Residential mobility is here apprehended through successive evolution in legislation that has covered protection against flooding for the last two centuries.

8.3.1. *The local shock of catastrophic flash floods*

Over the 1950–2011 period, the Gard region experienced three major floods. Two events in 1958 and 2002 had a severe impact on the whole department (more than 20 deaths and the equivalent of 15,000 euros' worth of damage in each case). The third, in 1988, caused a similar level of damage, but had a much smaller spatial hold since it only affected the city of Nîmes, the prefecture of the department (25% of the total population in the department).

The outlining of flood-prone areas – from which bans or restrictions in constructability come from – are defined through institutional mapping, controlled by the State, and called "*Plan de Prévention des Risques d'Inondation*" (PPRi – Flood Risk Prevention Plan)[7]. These maps are based on evaluation methods that were homogenized in Europe in 2007 by combining various considerations, such as the highest known water levels (seen during earlier events), with hydraulic modeling[8].

Figure 8.2 shows the evolution over time of the annual ratio between the average density of buildings in flood-prone areas and the average density of all buildings in the Gard department. We consider this ratio as an indicator of the increased awareness of the possibility of flooding, independent from the annual rate of construction, the latter varying greatly according to the economic considerations and the proportion of surfaces exposed to flooding.

First, we noticed the average level of this indicator, which acts as a reminder of the attraction of flood plains for urbanization [DUR 15b]. We then noted that, beyond the interannual fluctuations that remain limited, trends seem to be linked to the first two events, but never the third. We considered the decades following the three major events of 1958, 1988 and 2002 (excluding the 2 first years immediately following the event because of delays between decisions and declaration dates, as explained above). We added the linear regression line on the last seven points for periods after 1958 and 1988 (50% of variance explained). The two events of 1958 and 1988 have a visible effect on the indicator, with an incline of around one point per decade. This effect lasts around a decade and then more or less stabilizes itself, maintaining the decrease between the two peaks of the three decennial intervals. The 2002 event is in many ways comparable to that of 1958, yet it does not show any post-crisis effect. However, it should be noted that the proximity of this event to the end of the range used makes it difficult to truly assess a possible "stabilization".

7 For details on French flood legislation, see Chapter 5.

8 Directive framework (2007/60/EC) of the European Parliament and Council from October 23 2007 on the evaluation and management of flood risks OJ 288, 6.11.2007, pp. 27–34.

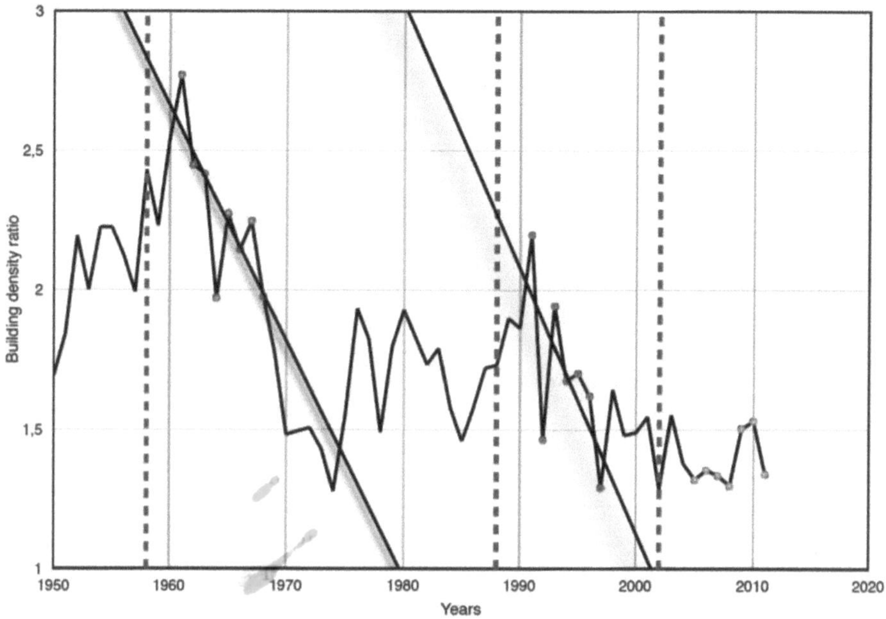

Figure 8.2. *Temporal evolution of the ratio between the average density of buildings in a flood-prone area and the average density of buildings in the Gard department (France). The dotted vertical lines represent the three last major events in the department (1958, 1986 and 2002). The regression lines are based on 10 years following the first events (excluding the first 2 years – see explanations on this point in section 8.3.1)*

The conclusions we can come to, thinking beyond the socioeconomic considerations of the property market that are not studied here, are that (1) catastrophic floods locally influence the building decisions for around a decade, and (2) this influence reaches its limit when the building conditions between flood-prone areas and non-flood areas balance out.

This result raises several questions. From a social point of view, it would be useful to better understand the mechanisms behind the observed post-1958 and 1988 event trends in order to assess their comparability, especially considering that, from a hydrometeorological point of view, these are two very distinct events (in terms of spatial spread). It would also be interesting to understand why lower thresholds are reached. From a physical point of view, it is also of interest to know the characteristics of events that are often qualified as "major" compared to other flash floods occurring over the same period but considered as "minor" since they resulted in a lot less damage and fewer victims, and often affecting smaller areas.

8.3.2. *The national evolution of the control of residential choices*

Chapter 5 examines the evolution of legislation in terms of flooding at the national level over more than two centuries (1807–2015), increasing the analysis to the European scale for the recent period. The base hypothesis is that there is an interaction between the evolution of legislation and residential mobility when faced with floods. On the one hand, institutional rules influence residential mobility – for the present case, this is a progressive change in attitude with regard to floods, going from resistance to acceptance. On the other hand, "national" catastrophic events highlight the problems linked to residential mobility during flooding, and in turn influence the regulatory framework as well as the social culture and "institution" [HOF 99]. Through this, we mean that the evolution of the law is linked to the individuals' representations of floods (culture) and the way in which they organize themselves to face them (social institution). Risk management is a complex problem due to the balance between potential losses and the level of investment needed to protect people, taking into account all the other development stakes. The solutions to this complex problem are necessarily taken conventionally and positioned historically. The prevailing selection criteria for the evolution of management texts cannot be defined based on efficacy alone (since this is always relative) but translate that on which an agreement is found – at least temporarily [BEC 14].

If we focus on the case of France, three main stages in the regulation of flood-prone areas can be distinguished (Table 8.2)[9]. An initial long period corresponds to the transition toward the idea that protection from floods is a public problem. This period goes from the 17th Century to the catastrophic national floods of the mid-19th Century. The two following periods mark the transition from a technicist view of active protection (1858–1935) to an anthropocentric view of passive protection (1935–2007), two periods separated by the catastrophic floods in the middle of the 20th Century. Finally, a fourth period (1964–2016) is identified to underline the recognition of the flooding problem – along with all other water-related problematics – as a multiscale challenge that, from an institutional point of view, deserve a correct balance in matters of subsidiary: floods develop at all levels and require institutional responses adapted to appropriate territorial and administrative scales.

9 The analysis of the evolution of legislation on matters of flooding, presented here in a synthesized manner for the sake of this chapter's arguments, is presented in detail in Chapter 5.

Date	Name of the law	Content of the law	Evolution of the governance paradigm	
1669	Royal ordinance on water and forests	First centralized law on water and forests that puts water mobility under royal protection; does not broach the subject of floods.	Emergence of the subject of flooding from other uses for water and the question of drying the land (1669–1858)	
1807	Law relating to the draining of marshes	Law on the drainage of wet areas, giving the control of water mobility to property owners in the affected areas. First French law explicitly referring to flooding.		
1854	Catastrophic floods of the *Rhône,* the *Loire* and the *Garonne*			
1856	Law on drainage	Law on agricultural drainage, voted hastily after catastrophic floods that led to an appeal for passive protection.		
1858	Law on the defense works of towns against flooding	In reaction to the 1854 flood, the first French law on protection against floods allowed the administration to build, modify or remove dikes in areas defined as flood prone.	Technicicst paradigm on the control of floods and river embankments (1858–1935)	
1898	Law on the water regime	Concerning the question of water ownership, this law does not take flooding into consideration.		
1910	Catastrophic floods of the *Seine* and the *Loire–Paris* floods			

1914	Bougère draft law	Interrupted by the 1st World War, this draft law favors passive protection (upstream dams), highlighting past management errors and advocating for the evaluation of private damage.		
1930	Catastrophic floods of the *Garonne* and the *Tarn*			
1935	Decree law on water flow	Limiting construction in submersible areas, first French law that subjects ownership to a public servitude, initiating the control of residential mobility.		
1955	Decree and application of article 91 of the urban development and housing code	Placing conditions on construction in risk areas, this decree lists the reasons to refuse building permits.		
1964	Law relative to the water regime and distribution, and the fight against their pollution	Addressing the problematic of water pollution, this law introduces basin agencies – administrative units based on the hydrographic structure	Anthropo-centric view of the control of residential mobility (1935–2007)	
1982	Law relative to the compensation of victims of natural disasters	Law on insurance for natural disasters that establishes the risk mitigation plans, consisting of cartographies and risk management rules.		Taking into account of the multiscale character of flood management (1964–2016)
1992	Law on water	Role of risk mitigation plans in the preservation, restoration and extension of flood plains.		
1999–2002	Catastrophic floods of the *Aude,* the *Hérault,* the *Gardon* and the *Tarn* and the Somme			

2007	Directive relative to the evaluation and management of flood risks	European directive coordinating the evaluation, mapping and management of floods at the European scale.		
2015	Law updating the territorial organization of the Republic	Transfer of the management of aquatic environments and the prevention of floods to intercommunities (intermediate territorial structures).		

Table 8.2. *Evolution of French legislation on matters of flooding over the last two centuries. The catastrophic events mentioned in the table are exclusively referred to as such in parliamentary debates during the voting for the cited laws*

A breakdown into three distinct periods is proposed for the analysis: in reality, the evolution is not outlined quite so simply and the influences can be felt from one period to another. The technicist paradigm therefore still persists in certain aspects today. These periods, nevertheless, convey what is seen over the course of the history of the problematics that dominated and influenced the evolution of legal texts.[10]

Analyzing the evolution of texts regulating the French management of floods underlines the catalytic effect of catastrophic events for the observed changes. Figure 8.3 represents the coevolution between events that are labeled "national" due to their magnitude, and the changes in the previously described policies.

This representation clearly highlights the shift from a management method arising from a problem of draining agricultural lands to active protection against flooding (period 1–2), and then from this management method of active protection to a more passive protection (period 3–4). We can therefore refer to an evolving progression with a characteristic timescale for each period of several decades. The analysis of the evolution of texts in the United States showed a comparable general trend, although with shorter timeframes (periods of around three decades) [TRU 14].

10 A comparable analysis – and more detailed – was done on legislation in the United Kingdom [PEN 06] providing a similar model of gradual change in policies that align well with the present analysis.

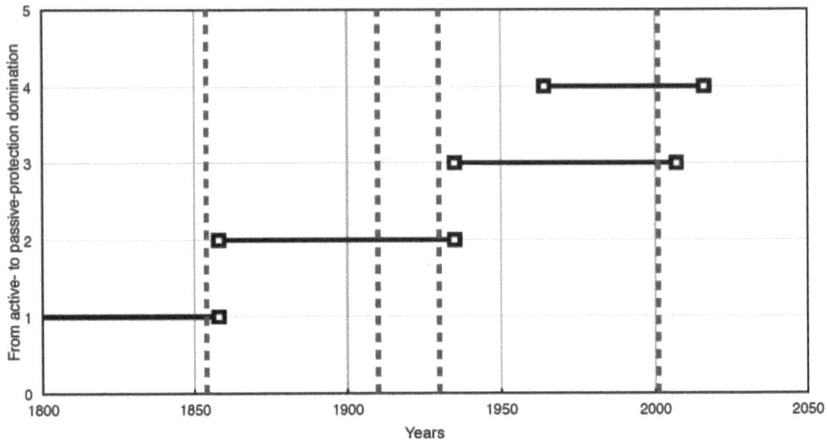

Figure 8.3. *Schematic illustration of the evolution of the institutional view of flood protection. Time is represented on the abscissa (over three centuries). On the Y axis, a graduation (arbitrarily on four levels) of active protection toward passive protection is represented. The vertical bars represent the "national" disasters explicitly mentioned during debates*

These results call for more research in order to refine the analysis. At the societal level, the analysis would benefit from a study of the different elements that influenced the evolutions; a more thorough study of the progression of concepts and technologies in the "water and forest" and "bridge and highway"[11] engineering trades, for example, would be highly instructive. At the physical level, it would be interesting to verify whether the so-called "national" flood events present common characteristics and whether their "elevation" – to put it this way – to the national level is only due to the social trauma caused, as intuition suggests. As we mentioned in Table 8.2, the trigger events for the 1856 or 1910 evolutions were catastrophic floods that were striking in their spatial magnitude, whereas those of 1930 or 2000 were spatially more limited but also striking due to their intensity and repetition. Our analysis does not shed sufficient light on the recent effects of the smaller scale floods that marked the end of the 20th Century and steered legislation toward a vision of multilevel management (period 4, visible in the last column of Table 8.2). The question of the subsidiarity that this change creates is not within the scope of this chapter; its analysis does, however, deserve the greatest attention[12].

---

11 These two trades that were historically distinct have been fused today to form the Bridge, Water and Forests Trade (IPEF).

12 The reader can refer to Chapter 5 in which this question is developed.

8.3.3. *The multiscale interactions that influence the transition from active to passive protection*

Following the two levels of analysis broached in the preceding sections (society-flash flood interactions at the local and national levels), we can underline, for the last century, a tendency to respond to major floods through a gradual constriction of residential mobility. The local response to floods is characterized by a decennial timescale, whereas its integration at the national level follows a more multidecennial rhythm. The characteristic amount of time to mark a notable difference between two laws is around seven decades in France[13].

In this section, we focused on the reaction time to floods at the local and national level, assuming that the same reasoning for the balance between passive and active protection is applicable to both levels. The national level shows a long and continuous passage from active protection to passive protection, marked by a few "national" floods. The local level changes more systematically and rapidly in reaction to "local" floods, but only for a relatively brief amount of time.

This result therefore appears counterintuitive in the sense that the national level covers larger territorial experiences, in which the number of extreme floods is greater than at the local level. This raises at least two questions. The first concerns the necessity of better understanding the nature of these "major floods"[14] that have the capacity to impact at a national level. The second concerns the role of the interscale links between the local and national levels that could provide a greater insight for understanding the reaction times shown.

8.4. The society–flash flood interactions across levels: post-crisis

After a crisis, in the weeks and months that follow the interactions between society and floods are under the dual influence of long- and short-term problems. The occurrence of floods and the manner in which they unfold then influence both residential mobility and daily mobility. Following the main trend of Table 8.1 and with recourse to the same data and methods as used in the analyses of sections 8.2 and 8.3, this section aims to examine the modifications that can appear in terms of human mobility following a flood, from two angles. The first angle is that of the decisions – sometimes radical – affecting the areas of residence; the second is that of the measures taken to adapt, in a more long-term manner, daily mobility in a sector regularly subject to flooding.

---

13 It is three decades in the United Kingdom [DEN 06].

14 On this point, see Chapter 4.

8.4.1. *Residential mobility in reaction to the damage caused by a flash flood*

After a flash flood, the damage caused on the impacted buildings and territories, as well as the trauma experienced by the individuals, generates decisions that are variable case by case with respect to the area of residence. The flood leads to a change in status for people, going from "exposed" to "victim" status. These victims are variably impacted based on anterior choices that were made on matters of human mobility in the area – covered in the previous section. The reactions to affected buildings, whether housing, commercial or industrial establishments, or even public buildings, are filled with emotion and cannot be summarized by the simple decision to repair or abandon the location.

An example is provided by the case study in the village of Collias – which has since become a case study for matters of French expropriation permitted by the Barnier law. It is a small village (around 20 km^2 and 1,000 inhabitants) located at the confluence of two rivers: the Gardon and its tributary, the Alzon. The catastrophic flood of 2002 – which resulted, at this location, in the unfortunate concomitance of peak discharges of the two rivers – heavily impacted 95 residences, and even completely destroyed some of them. Thanks to the intuition and vigilance of the mayor at the time, the anticipated evacuation of inhabitants ensured that there were no victims to mourn. In response to this event, local authorities, accompanied by State services, inaugurated the possibility provided by the recent Barnier Law (1995) to compensate the most exposed inhabitants. The decision was made to apply the expropriation procedure intended by the law[15] and to delocalize all affected buildings into a zone exempt of risk [JOU 16]. Interviews[16] carried out with the mayor responsible in 2002, as well as with 14 of the village inhabitants – of which several were directly affected by the expropriation – allowed a measure of the difficulties, numerous and complex, related to following such a decision to be taken. Some inhabitants, shocked by the magnitude of the event and the damage done to their living space, easily accepted the offer of compensation for their goods and the delocalization plan. They had new houses built, in which they have since lived, sheltered from floods. Others, however, could not accept this idea of abandoning their goods and took up a position of resistance against the will of local and national authorities. Some still fight against the State today, more than 15 years after the flood, engaged in morally and financially long and costly legal battles. Moreover, this example is also striking for the manner in which the mayor of Collias managed the exposure through measures of acceptance, whether in the short term of the crisis or the longer prevention time. On the basis of

---

15 The law effectively stipulates the possibility of an expropriation imposed by the State in certain conditions of extreme danger.

16 Interviews carried out in the context of a socioethnographic study for the Mobiclimex project, see Chapter 2 and Appendix 2.

earlier experiences of minor floods (1994–2001), she realized that the regional warning system was not pertinent and would arrive too late (if at all) to the village of Collias. Consequently, in 2001, she established a local evacuation system that she fittingly activated in 2002, foreseeing the evacuation of exposed residents as well as people involved in a trade show located in the enclaved area.

The case of displacement from Collias is exceptional compared to the general trend seen in Gard with regard to residential mobility[17]. Although flooding does not appear as a dominant factor in deciding to move for most of the interviewed inhabitants, there are a few examples for which the violence of the flood was enough to scare the individuals – often new arrivals in the south east of France – to sell their house and leave the flooded area, even the region. We could also mention the case of the 1992 flood in Vaison-la-Romaine (France), which dramatically affected four recently built neighborhoods on the border of the river (the Ouvèze). About 20 houses were carried away by the very violent flash flood and 12 deaths were reported [COM 93].

Most of the time, however, floods are not sufficient motive to leave a place of residence. In regions that face regular flooding, floods of smaller magnitude – qualified as "regular" – are more likely to encourage active protection responses (construction of dikes, for example). These measures can produce perverse effects within the riparian population, pushing them to feel protected and not adopt the recommended preventative behaviors [DUR 15b]. Protection measures effectively influence the perception of risks [MCP 77] and the recollection of events can gradually become diluted through the adoption of successive protection measures [NOV 11]. This can generate a belief according to which the work reduces the risks to zero [PIN 05] and can lead to populations no longer being evacuated for major floods, as illustrated by the dramatic example of the Biot (France) flash flood in October 2015. A resulting four victims were mourned from a nursing home in which the water rose to the first floor, whereas the decision to evacuate was not taken due to the protective wall that had just been built after several smaller floods. More broadly, spaces that are protected with active protection measures are more easily accompanied by an attitude of resistance. This phenomenon is encouraged by the fact that active protections rightly remove the incidence of minor floods [BAU 13], thus giving rise to a sense of dominance over the elements [DUR 14].

8.4.2. *The preparation of daily mobility for future floods*

Mobility models are known for being designed over a wide range of timescales. They are modified according to lifestyle evolutions (in the long term), land

17 Survey led in Gard, in the context of the MobiClimEx project, on residential trajectories (see Chapter 2 and Appendix 2).

management decisions (in the medium term), as well as in individual choices of residence and work location (in the medium and short terms), inducing annual to daily adaptations. In the very short term, daily mobility decisions are finer and impulsive or opportunistic depending on the more or less unpredictable constraints that are both internal (such as a sick child who cannot go to school) and external (road works, for example) [SAL 83, DOH 00, BEI 08, VAN 10]. The occurrence of a flood, whether extreme or not, is often the occasion to take initiatives that could be called corrective on daily mobility.

An example of a collective response made after several events to prepare for future "extreme" floods is the "Plan POTES" procedure (*Plan de l'organisation des transports et des établissements scolaires lors d'événements climatiques* – "organization of transport and school establishments during climatic events plan"). This plan was tested and adopted in the Gard prefecture shortly after the 2002 disaster to manage schools and to organize school bus services during climatic events. The last crisis showed that the safety of students was one of the main mobility motivations of parents during an event, and thus a source of endangerment. The head of the emergency management services department therefore considered the mobility of students as a major collective preoccupation. The POTES plan defines the crucial moment before which central security decisions, such as the anticipated evacuation of schools, must be taken and announced to prevent chaotic situations in which school buses and/or parents find themselves on flooded roads. A calendar was defined based on the understanding of family routines, school recommendations and the constraints of the school bus service, taking into account the delay necessary for informing all these stakeholders. The design of the procedure came from a collective process that allowed the different players to meet, to understand and assess each other's preoccupations and constraints, and to communally define certain regulations for sharing the responsibilities and actions during the crisis [RUI 07a].

The "ordinary" floods therefore forced certain human actions that accumulated over time to constitute a "common base" of knowledge in preparation for future occurrences. On the cumulative learning curve with regard to floods, the departmental services of Gard constantly work on the roads during events[18]. Gradually, an intricate knowledge is accumulated, enabling the management of protection on roads, whether through the construction of water roads, the modification of routes (active protection and a form of resistance) or the proposal of alternative paths when roads are flooded (passive protection and a form of acceptance).

18 A detailed inventory of the roads in Gard was therefore created over the last 40 years [VER 10b, VER 10b].

8.4.3. *Lessons learned through floods*

These examples of management through the cumulative knowledge acquired from occurrences illustrates how individuals and collectives learn from a crisis and attempt to adapt, notably by modifying their residential and daily mobility. Each flood highlights the weaknesses of prevention systems and pushes for the consideration of improvements for the future. The crisis also imposes a form of reflection on the interactions between water mobility and the mobility of man. As such, the post-crisis period is a precious time for the establishment of new prevention procedures. Specific security plans for school establishments (PPMS) and familial safety plans (PFMS) at the family level were thus set up. Since 2002 and 2004, these plans have obliged school directors and families to formalize their information and protection methods in preparation of future natural or technological events[19]. In this context, the recommendation with regard to daily mobility is simple: avoid movements and do not go and pick up children at school.

These procedures plot the transition toward the subsidiarity of the French risk management addressed in section 8.3. They recognize that individuals must face local situations and that these must be managed case by case. This transfer of responsibility from the State to local authorities and citizens to ensure their own protection has become the norm, in a context of increasing power behind the idea of resilience when faced with risks, in France as well as elsewhere [REG 10]. Even if these actions are in place at the public relations establishment level – especially when the public are young – and encouraged at the individual level, by sensitization campaigns for example, a weakness still remains due to the absence of an equivalent within the business sector. Although private enterprises, just like public establishments, have the responsibility of ensuring the safety of their employees within the framework of their current activities, there is no form of established regulation for the management of these "out of the ordinary" moments during which their own resources must be mobilized to face the climatic extremes and floods. Here, all evidence suggests that there remains a field of work to be implemented.

8.5. Conclusion: multiscale considerations

In this chapter, we examined the interactions at work in the different levels between water mobility and human mobility with regard to flash floods, while taking into consideration the wide categories of variability for the long and short

19 Although the establishment of the PPMS was able to be applied in schools by state services and are today systematic in exposed areas, the PFMS – with no equivalent pressure tactic – did not have the same diffusion success and until today remain relatively little known by families.

term, as summarized in Table 8.1. Throughout the three preceding sections, we provided descriptions and references documenting the typical frequency for the mobility of water and people. We illustrated our statements on these short-term (Figure 8.1) and long-term (Figures 8.2 and 8.3) interactions through examples and data from our successive studies in south east France. Figure 8.4 sketches a compilation of these frequencies that characterize the issue of floods from a sociohydrological point of view. This multiscale description opens up perspectives for future research with operational implications.

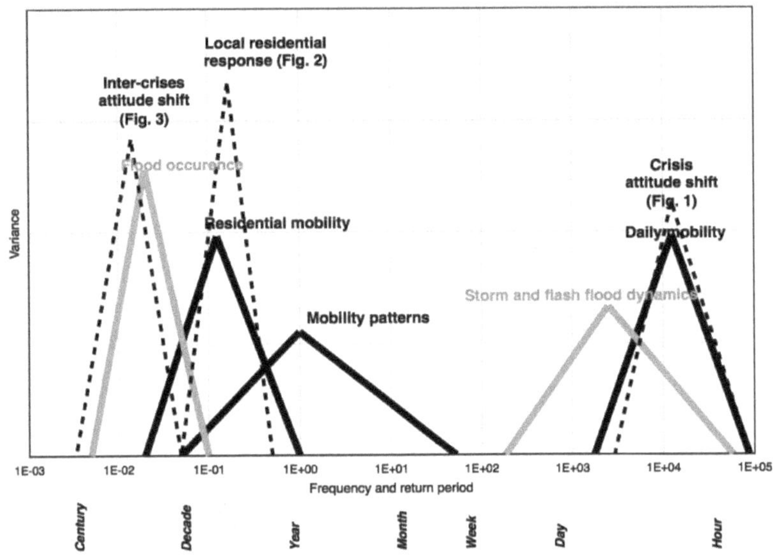

Figure 8.4. *Schematic representation of the spectrum of temporal variability of the human mobility and water mobility processes. The variance axe (Y axis) is arbitrary, we have simply defined the surfaces equal to the different graphs, which is logical from a mathematical point of view and facilitates its reading. The curves in bold represent the characteristic frequencies of water and human mobility taken from the reference cited in the text. The dotted curves represent the characteristic frequencies of three water–society interactions illustrated in Figures 8.1–8.3*

8.5.1. *The institutional response to floods (by flash flooding) in time and space*

From a strictly hydrometeorological point of view, there is a very clear separation of timescales between the appearance of extreme local storms triggering catastrophic floods and the development dynamics of extreme phenomena and flash floods (gray curves in Figure 8.4). The physical mechanisms relative to the local occurrence of floods are quite well analyzed and have begun to be linked to human mobility dynamics in several studies (of which some are cited in the preceding sections). However, flash floods development dynamics in the longer term require renewed attention because, if we are interested in the long-term societal response to these phenomena, the notion of an "impact event" takes on a very different meaning at the local level compared to at the national level (as shown in section 8.3).

Institutional regulation and norms evolve differently under the pressure of these "impactful events", of rare occurrence, over more or less large territories depending on the institutional level in question (Table 8.2 and Figures 8.2 and 8.3). On a small territory, we refer to events that generally have a return period of around 50 years (see the occurrence of floods outlined in Figure 8.4). Yet even at this scale, attributing a return period to an event is a separate research subject (on this point, see how [RAM 05] attempts to address this question through the notion of gravity). On a much vaster territory covering several river systems, typically at the country scale, understanding what defines an "impactful event" is even more delicate. The institutional response rate at this level is obviously not that of the sum of events impacting at the local level. According to the analysis of the evolution of legislation summarized in Table 8.2, we saw that the national institutional rhythm is marked by the occurrence of severe floods in different places, arising either from a common meteorological event, or from different events accumulated over a small period of time (within a year).

These different observations deserve new investigations. Beyond the many other influencing factors (notably economic and political), severe floods influence in an obvious way, at least partially, the cascade of institutional decisions and actions taken between the national and local levels. The multilevel sociotechnical pathway models [GEE 07] could help to clarify the way in which these interactions function and to better understand where the processes in play occur, as well as the characteristic timeframes behind this institutional cascade game.

The recurrence of major floods over vast spaces is the result of meteorological forcing on a large scale from favorable local conditions to the development of extreme events and specific social exposure and vulnerability. *Event attribution studies* aim to separate the large-scale forcing influence (dynamics) from the small-scale development of clouds and local precipitations (thermodynamics), providing an example of the kinds of studies that can be led on this point [SCH 16, VAU 16]. Such an approach would allow for a better understanding of concurrence conditions at the large scale and their potential evolution through climate change. It could also invite improved descriptions of the random character of atmospheric thermodynamics at the local scale and their resonance with hydrology.

From an operational point of view, a better understanding of the "extreme conjunction of extremes" or even the "ordinary succession of extremes" could be very useful for answering the range of questions asked (and that will certainly be asked more and more) on the recurrence of major floods in a context of climate change in terms of planning, public policies and, more broadly, society.

8.5.2. *The adaptation of human mobility models at the intermediate scale*

The range of scales on which questions of human mobility are asked largely cover the hydrometeorological scales examined above (Figure 8.4). As shown in Figure 8.2 and Table 8.2, the characteristic response time of national regulation on matters of flooding depends on the occurrence of events at the local scale. When looking at Figures 8.1 and 8.3, we see that the responses in terms of mobility when faced with flood dynamics (long term), on the one hand, and the occurrence of a flood (short term), on the other hand, find themselves, respectively, in the same order of time as the characteristic time of residential mobility and daily mobility. In Figure 8.4, it appears that, although residential mobility is staggered by an order of magnitude compared to the dynamics of floods at the local scale (in an order of several decades against half a decade for residential mobility), the evolution of daily mobility plans (or patrons) adds a new, intermediate level of interest, in the order of a year.

This notion of a mobility plan, used to link residential mobility and daily mobility, deserves particular attention in terms of flood risk management. The way in which residential mobility governs daily mobility through these plans is not just a theoretical concept, but appears possible today because of new methods of following the position of people in time, notably through the use of mobile phones. Gonzalez *et al.* [GON 08] showed that these mobility plans are very regular in time and space, with the presence of certain highly frequented poles of attraction. To our knowledge, most mobility studies on emergency interventions concentrate on crisis and post-crisis mobility at scales which we call daily (Figure 8.4). With regard to this, mobile phone data seems promising for providing detail on the observations [GET 11, SON 13]. The infra-weekly and infra-daily scales studied for transport problems could help improve the prediction of extreme meteorological phenomena and warning systems, by allowing for a better understanding of human exposure based on the time of day, or from one day to another [CRE 09, DEB 16, SHA 17].

For the intervals ranging between days and years that we call patrons of mobility in Figure 8.4, we postulate that extreme meteorological conditions and "ordinary" floods can have an influence on these mobility habits that vary depending on weekly or seasonal activities (alternation of working days, weekends, working periods, holidays, etc.). Among the people questioned in the department of Gard, many mentioned that they know where road traffic was usually blocked during heavy rains and that they had other, safer routes to use at these times [RUI 07b]. This leads us to consider that a better understanding of the role of "ordinary" bad weather conditions on the adaptation of these patrons of mobility is necessary.

From an operational point of view, the succession of "ordinary extremes" could constitute opportunities to motivate the adaptation of these daily mobility practices. The idea is to communicate on the manner of treating the (extraordinary) extremes by leaning on these "ordinary extremes". As previously mentioned, the "active" or structural measures that limit the appearance of these "ordinary extremes" deprive societies of the opportunity of such a learning experience [BAU 13, DUR 14].

8.5.3. *The multiscale alternation from active protection to passive protection, in time and space*

From the point of view of human behavior, this chapter distinguishes two main attitudes in terms of mobility during floods: resisting by confronting the power of the water or accepting to remove oneself from the flood by moving (to a secure location if referring to the short term or to a relocation in terms of mobility in the long term). These two attitudes can be seen at different levels of mass [GRO 06][20], from the individual reaction to the collective response at the country level (such as the national policy on flood areas, or the "Barnier" fund which allows for expropriation for example).

We hypothesize that there is a general evolving trend from an attitude of resistance to events to their acceptance (Figures 8.1 and 8.3), but that there can be an alternation in time between these two behaviors (Figure 8.2). We developed examples such as that of the mayor of Collias in Gard who maintained the same attitude in the short term and long term. This description is, unfortunately, too partial to give a consolidated view of this alternation. However, such a view would allow, on the one hand, the description of the human factors that underpin these attitudes, and, on the other hand, a better understanding of the most appropriate temporal methods and windows for communicating the best ways of adapting one's behavior when facing a flood risk.

After a long period dominated by a technocentered view of floods, giving the illusion that centralized institutional resolutions to problems were possible, we now note an evolution toward a more distributed view. Recent subsidiarity, which is

20 The notion of a "mass scale" refers to Grossetti and the number of interactions between individuals, defining the levels of mass considered.

slowly imposing itself in flood management matters, aims to find the best way to resist or adapt, and is variably based on the considered scale: mass social scale (individual practices, group interactions within neighborhoods, community initiatives, institutional rules), temporal scale (crisis, postcrisis and intercrisis) or "level of generality"[21] [GRO 06] – from decisions specifically taken to face flooding to the set of protection methods integrated in life's daily activities. This is what leads us to believe in the importance of developing and teaching this multiscale approach to flood management, especially for future managers [RUI 12].

8.6. Bibliography

[BAU 13] BAUDUCEAU N., "Éléments d'analyse sur les ressorts de la localisation des habitants en zone inondable", *Séminaire MobiClimEx*, Ecoles des Mines, Paris, September 13, 2013.

[BEC 14] BECKER H., *What About Mozart? What About Murder?*, University Chicago Press, Chicago, 2014.

[BEI 08] BEIGE S., AXHAUSEN K.W., "Long-term and mid-term mobility decisions during the life course – experiences with a retrospective survey", *IATSS Research*, vol. 32, no. 2, pp. 16–33, 2008.

[BLO 95] BLÖSCHL G., SIVAPALAN M., "Scale issues in hydrological modelling: a review", *Hydrological Processes*, vol. 9, pp. 251–290, 1995.

[BOR 08] BORGA M., GAUME E., CREUTIN J.-D. *et al.*, "Surveying flash floods: gauging the ungauged extremes", *Hydrological Processes*, vol. 22, pp. 3883–3885, 2008.

[BOR 11] BORGA M., ANAGNOSTOU E., BLOESCHL G. *et al.*, "Flash flood forecasting, warning and risk management: the HYDRATE Project", *Environmental Science and Policy*, vol. 14, pp. 834–844, 2011.

[CER 10] CERESETTI D., MOLINIÉ G., CREUTIN J.-D., "Scaling properties of heavy rainfall at short duration: a regional analysis", *Water Resources Research*, vol. 46, pp. 12, 2010.

[CER 11] CERESETTI D., ANQUETIN S., MOLINIÉ G. *et al.*, "Multiscale evaluation of extreme rainfall event predictions using severity diagrams", *Weather and Forecasting*, vol. 27, pp. 174–188, 2011.

[COM 93] COMBY J., "Bilan social et économique de la séquence orageuse du 22 septembre 1992 dans le sud du couloir rhodanien", *Revue de géographie de Lyon*, vol. 68, pp. 175–192, 1993.

[CRE 09] CREUTIN J.-D., BORGA M., LUTOFF C. *et al.*, "Catchment dynamics and social response during flash floods: the potential of radar rainfall monitoring for warning procedures", *Meteorological Applications*, vol. 16, pp. 115–125, 2009.

21 Through this term, Grossetti [GRO 06] refers to the level of specificity of a situation.

[CRE 13] CREUTIN J.-D., BORGA M., GRUNTFEST E. *et al.*, "A space and time framework for analyzing human anticipation of flash floods", *Journal of Hydrology*, vol. 482, pp. 14–24, 2013.

[DEB 16] DEBIONNE S., RUIN I., SHABOU S. *et al.*, "Assessment of commuters' daily exposure to flash flooding over the roads of the Gard region, France", *Journal of Hydrology*, vol. 541, pp. 636–648, 2016.

[DEL 05] DELRIEU G., NICOL J., YATES E. *et al.*, "The catastrophic flash-flood event of 8–9 September 2002 in the Gard Region, France: a first case study for the Cévennes–Vivarais Mediterranean Hydrometeorological Observatory", *Journal of Hydrometeorology*, vol. 6, no. 1, pp. 34–52, 2005.

[DIB 13] DI BALDASSARRE G., VIGLIONE A., CARR G. *et al.*, "Socio-hydrology: conceptualising human-flood interactions", *Hydrology and Earth System Sciences*, vol. 17, pp. 3295–3303, 2013.

[DIB 15] DI BALDASSARRE G., VIGLIONE A., CARR G. *et al.*, "Debates—perspectives on socio-hydrology: capturing feedbacks between physical and social processes", *Water Resources Research*, vol. 51, pp. 4770–4781, 2015.

[DOH 00] DOHERTY S.T., MILLER E.J., "A computerized household activity scheduling survey", *Transportation,* vol. 27, no. 1, pp. 75–97, 2000.

[DUR 14] DURAND S., Vivre avec la possibilité d'une inondation ? Ethnographie de l'habiter en milieu exposé… et prisé, PhD Thesis, Aix Marseille University, France, 2014.

[DUR 15a] DURAND S., "Du marécage à la zone bétonnée et protégée: Trajectoire d'une zone inondable", *Les annales de la recherche urbaine*, vol. 110, Ville et vulnérabilités, September 2015.

[DUR 15b] DURAND S., "Une réponse technique au devoir d'information et de sensibilisation. De la difficulté à communiquer sur le risque résiduel", *Colloque de la Société hydrotechnique de France "Gestion des risques d'inondation",* Paris, May 27 and 28, 2015.

[GAU 04] GAUME E., LIVET M., DESBORDES M. *et al.*, "Hydrological analysis of the river Aude, France, flash flood on 12 and 13 November 1999", *Journal of Hydrology*, vol. 286, pp. 135–154, 2004.

[GEE 07] GEELS F.W., SCHOT J., "Typology of socio-technical transition pathways", *Research Policy*, vol. 36, pp. 399–417, 2007.

[GET 11] GETHING P.W., TATEM A.J., "Can mobile phone data improve emergency response to natural disasters?", *PLoS Medicine*, vol. 8, pp. e1001085, 2011.

[GON 08] GONZALEZ M.C., HIDALGO C.A., BARABASI A.-L., "Understanding individual human mobility patterns", *Nature*, vol. 453, pp. 779–782, 2008.

[GRO 06] GROSSETTI M., "Trois échelles d'action et d'analyse", *L'année Sociologique*, vol. 56, no. 2, pp. 285–307, 2006.

[HOF 99] HOFFMAN A.J., "Institutional evolution and change: environmentalism and the US chemical industry", *Academy of Management Journal*, vol. 42, pp. 351–371, 1999.

[JOU 16] JOUANNIC G., CROZIER D., CHLOÉ T.D.M. *et al.*, "Post-disaster recovery: how to encourage the emergency of economic and social dynamics to improve resilience?", *E3S Web Conference*, vol. 7, p. 17004, 2016.

[LAN 13] LANG M., COEUR D., BARD A. *et al.*, "Les inondations remarquables en France: premiers éléments issus de l'enquête EPRI 2011", *La Houille Blanche*, vol. 5, pp. 37–47, 2013.

[LIU 01] LIU J., DIETZ T., CARPENTER S.R. *et al.*, "Complexity of coupled human and natural systems", *Science*, vol. 317, no. 5844, pp. 1513–1516, 2001.

[LUT 16] LUTOFF C., CREUTIN J.-D., RUIN I. *et al.*, "Anticipating flash floods: multi-scale aspects of the social response", *Journal of Hydrology*, vol. 541, pp. 626–635, 2016.

[MAR 10] MARCHI L., BORGA M., PRECISO E. *et al.*, "Characterisation of selected extreme flash floods in Europe and implications for flood risk management", *Journal of Hydrology*, vol. 394, pp. 118–133, 2010.

[MCP 77] MCPHERSON H.J., SAARIEN T.F., "Flood plain dweller's perception of the flood hazard in Tucson, Arizona", *The Annals of Regional Sciences*, vol. 11, no. 25, pp. 25–40, 1977.

[MOL 12] MOLINIÉ G., CERESETTI D., ANQUETIN S. *et al.*, "Rainfall regimes in a Mediterranean mountainous region", *Journal of Climate and Applied Meteorology*, vol. 51, pp. 429–448, 2012.

[NOV 11] NOVEMBER V., PENELAS M., VIOT P. (eds), *Habiter les territoires à risques,* Presses polytechniques et universitaires romandes, Lausanne, 2011.

[NUI 08] NUISSIER O., DUCROCQ V., RICARD D. *et al.*, "A numerical study of three catastrophic precipitating events over southern France. I: Numerical framework and synoptic ingredients", *Quarterly Journal of the Royal Meteorological Society*, vol. 134, pp. 111–130, 2008.

[NUI 09] NUISSIER O., JOLY B., JOLY A. *et al.*, "A statistical downscaling to identify the large scale circulation patterns associated with heavy precipitation events over southern France", *Quarterly Journal of the Royal Meteorological Society*, vol. 137, pp. 1812–1827, 2009.

[PEN 06] PENNING-ROWSELL E., JOHNSON C., TUNSTALL S., "'Signals' from pre-crisis discourse: lessons from UK flooding for global environmental policy change?", *Global Environmental Change*, vol. 16, pp. 323–339, 2006.

[PIC 88] PICON B., *L'espace et le temps en Camargue,* Actes Sud, Arles, 1988.

[PIN 05] PINTER N., "One step forward, two steps back on U.S Floodplains", *Science*, vol. 308, no. 5719, pp. 207–208, 2005.

[PLA 08] PLAN BLEU, Changement climatique et énergie en Méditerranée, Sophia Antipolis, 2008.

[RAM 05] RAMOS M.H., CREUTIN J.-D., LEBLOIS E., "Visualization of storm severity", *Journal of Hydrology*, vol. 315, pp. 295–307, 2005.

[REG 10] REGHEZZA-ZITT M., "La résilience dans les politiques françaises de gestion des inondations urbaines: quelques pistes d'analyse", *Bulletin de l'Association de géographes français*, vol. 87, no. 4, pp. 503–516, 2010.

[RUI 07a] RUIN I., Conduite à contre-courant. Les pratiques de mobilité dans le Gard: Facteur de vulnérabilité aux crues rapides, PhD Thesis, Joseph Fourier University Grenoble I, France, 2007.

[RUI 07b] RUIN I., GAILLARD J.-C., LUTOFF C., "How to get there? Assessing motorists' flash flood risk perception on daily itineraries", *Environmental Hazards*, vol. 7, pp. 235–244, 2007.

[RUI 08] RUIN I., CREUTIN J.-D., ANQUETIN S. *et al.*, "Human exposure to flash floods – relation between flood parameters and human vulnerability during a storm of September 2002 in Southern France", *Journal of Hydrology*, vol. 361, pp. 199–213, 2008.

[RUI 09] RUIN I., CREUTIN J.-D., ANQUETIN S. *et al.*, "Human vulnerability to flash floods: addressing physical exposure and behavioral questions", in SAMUELS P. (ed.), *Flood Risk Management: Research and Practice*, Taylor and Francis Group, London, 2009.

[RUI 12] RUIN I., LUTOFF C., CRETON-CAZANAVE L. *et al.*, "Toward a space–time framework for integrated water and society studies", *Bulletin of the American Meteorological Society*, vol. 93, no. 10, pp. ES89–ES91, 2012.

[RUI 14] RUIN I., LUTOFF C., BOUDEVILLAIN B. *et al.*, "Social and hydrological responses to extreme precipitations: an interdisciplinary strategy for postflood investigation", *Weather, Climate, and Society*, vol. 6, no. 1, pp. 135–153, 2014.

[SAL 83] SALOMON I., BEN-AKIVA M., "The use of the life-style concept in travel demand models", *Environment and Planning A*, vol. 15, no. 5, pp. 623–638, 1983.

[SCH 16] SCHALLER N., KAY A.L., LAMB R. *et al.*, "Human influence on climate in the 2014 southern England winter floods and their impacts", *Nature Climate Change*, vol. 6, no. 6, pp. 627–634, 2016.

[SHA 17] SHABOU S., RUIN I., LUTOFF C. *et al.*, "MobRISK: a model for assessing the exposure of road users to flash flood events", *Natural Hazards and Earth System Sciences*, vol. 17, no. 9, 2017.

[SIV 12] SIVAPALAN M., SAVENIJE H.H.G., BLÖSCHL G., "Socio-hydrology: A new science of people and water", *Hydrological Processes*, vol. 26, no. 8, pp. 1270–1276, 2012.

[SIV 15] SIVAPALAN M., BLÖSCHL G., "Time scale interactions and the coevolution of humans and water", *Water Resources Research*, vol. 51, pp. 6988–7022, 2015.

[SON 13] SONG X., ZHANG Q., SEKIMOTO Y. *et al.*, "Intelligent system for human behavior analysis and reasoning following large-scale disasters", *IEEE Intelligent Systems*, vol. 28, pp. 35–42, 2013.

[TER 15] TERTI G., RUIN I., ANQUETIN S. *et al.*, "Dynamic vulnerability factors for impact-based flash flood prediction", *Natural Hazards*, vol. 79, pp. 1481–1497, 2015.

[VAN 10] VAN ACKER V., VAN WEE B., WITLOX F., "When transport geography meets social psychology: toward a conceptual model of travel behaviour", *Transport Reviews*, vol. 30, no. 2, pp. 219–240, 2010.

[VAN 96] VAN DINE D., Debris flow control structures for forest engineering, British Columbia Ministry of Forests, Working Paper, Research Branch, 1996.

[VAU 16] VAUTARD R., YIOU P., OTTO F. *et al.*, "Attribution of human-induced dynamical and thermodynamical contributions in extreme weather events", *Environmental Research Letters*, vol. 11, pp. 114009, 2016.

[VER 10a] VERSINI P.A., GAUME E., ANDRIEU H., "Application of a distributed hydrological model to the design of a road inundation warning system for flash flood prone areas", *Natural Hazards and Earth System Sciences*, vol. 10, pp. 805–817, 2010.

[VER 10b] VERSINI P.A., GAUME E., ANDRIEU H., "Assessment of the susceptibility of roads to flooding based on geographical information – test in a flash flood prone area (the Gard region, France)", *Natural Hazards and Earth System Sciences*, vol. 10, pp. 793–803, 2010.

Conclusion

This book provides an overview of the works of research that have been carried out over the past 15 years by an *interdisciplinary team* that has gathered geoscience and human and social science specialists together to better understand the *interaction between the dynamics of flash flood phenomena and social mobility phenomena*. The aim of these works is to provide the means to discern what happens in critical situations, considered as one of the manifestations of this interaction. This topic is all the more crucial today given the worry that such situations might multiply in the context of climate change.

We hypothesized that the *notion of scale* constitutes an interesting entry point for working on the different issues identified. In this conclusion, we propose a return to the main contributions of this notion and the identification of research paths that this reflection will help open.

Flash floods are phenomena that involve particular *spatial and temporal scales*. Assessing the interactions between these specific processes and social dynamics therefore involves, first of all, mobilizing these categories of scale and observing the social phenomena at play in the short term (several hours) and over restricted areas (from small-scale catchment basins to several thousands of km^2). At these constricted scales, one of the questions that remain is that of predicting these flash flood phenomena. As shown in Chapter 3, although progress has been made on improving the precision of local predictions, there are still fields of investigation to refine and deepen anticipation capacities. On the social side, questioning the time and space of a flood raises considerations of the social processes that they trigger and, based on this, the questioning of other categories of scale.

Chapter written by Céline LUTOFF and Séverine DURAND.

The impact of floods can take on very different forms at different *social scales,* from the individual to the State (even to international organizations), via collectives of variable geometry.

At the individual level, Chapter 2 highlights the importance of the *power of the daily routine* in decision-making, in terms of mobility ranging from the daily to the residential. Even though a specific event can have a strong impact on individuals at the moment of the crisis, it appears that their choice of residence is not heavily influenced by the risk. Where there is a strong attachment to the area lived in, floods appear as part of the inherent constraints of the area. The cross-analysis of the social effects of flooding over the short time span of the flood and the long timespan of life histories thus highlights the power of the daily routine, the anthropological disposition of maintaining the ordinary course or of scheduled activities regardless of the circumstances.

In this context, living with sometimes extreme variations in nature raises the major problem of *resistance to change,* already widely observed and worked on in different domains. This inclination to resistance is worrying in the face of possible multiplications of these events in a context of climate change.

This resistance to change is observed in both individuals and collectives. Chapters 5 and 6 therefore show how taking these phenomena and the required protection of property and people into account often comes up against resistance and constraints that raise other parameters. As such, *large evolutions in territory*, notably linked to economic transformations, can encourage the occupation of flood-prone areas. This is what happened in certain sectors of the Gard department, with, on the one hand, the development of tourism and jobs close to the coastal border, and, on the other, a decline in older activities, such as mining or sericulture in the Cévennes. These changes brought along a thorough evolution in the structure of the territory and the continued urbanization of certain sectors that are, however, susceptible to flooding.

However, when we move away from the *institutional scale* and look at the manner in which protection against flooding has been carried out throughout the history of norms and laws, as proposed in Chapter 5, we become aware of a certain evolution since the 19th Century. Although the process is slow, it is no less real. Certain local episodes were sufficiently striking to generate changes in the existing judicial tools. Regulations therefore mark an *evolution in conception of the standards*. Over the course of time, we see a progressive move from a conception centered on the uses of water and the shaping of rivers to enable the continuity of these uses to a raised awareness of the necessary collective management of flood protection. Beyond these, the increased awareness also gave rise to a greater evolution of the relationship with the flood. Initially leaning on the willingness to master the phenomena through knowledge and control of water flow, the successive standards emphasize the gradual consideration of the notion of exposure, of stakes and of vulnerability, guiding regulatory measures toward limiting human settlement in susceptible areas. This dynamic of taking flood risk into account was accompanied by a reflection on the adequacy of institutional scales, leading to exceedance of the traditional administrative limits. Today, this process is in a new stage following the 2014 promulgation of the MAPTAM law and the establishment of the GEMAPI jurisdiction in 2018.

Through this reflection of the levels pertinent to the analysis of interactions between flash floods and social mobility, this book also highlights the need to create new reference levels to address certain points that have been identified as major.

This notably concerns taking into account the potential impact of events. Chapter 3 thus concludes that today it is essential to go further than solely predicting outflows, in order to integrate their potential impact on society, infrastructures and human activities. Chapter 4 brings certain elements of a response in terms of the tools necessary for the construction of this scale of impact. Notably, it identifies the need for a database integrating both hydrometeorological observations and data concerning the social, economic and environmental impact, among others. The authors of this chapter then propose to aim for the development of a *severity scale* to respond, at least in part, to this need for the social consequences of a flood to be taken into account. However, Chapter 8 shows that the construction of this scale calls for further developments in specific research.

Beyond this notion of severity, Chapter 8 also opens up the consideration of taking into account levels of adaptation or resistance, observable depending on the contexts and the levels of action or decision considered. In this case, it involves evaluating the capacity to "live with" or, on the contrary, to "fight against" the flood. This chapter sheds light on the attitudes characteristic of each of these positions. Based on their observations, the authors hypothesize that there is a general trend toward "living with", but they also highlight a possible alternation between the two attitudes through the timeline of the event (precrisis, crisis, postcrisis), depending on the social scale and the level of generalities considered (specific situation or integrated management of the risk).

Progression in the analysis of these positions and attitudes notably involves questioning the rhythms and cycles of adaptation. A central issue consists of identifying which elements permit the switch from one reference model to another. Can extreme floods trigger profound changes in social dynamics and punctuate adaptation in a specific way for each level (spatial, social, institutional)? We propose exploring this question in the next volume, focusing once more on the work undertaken in southern France and the Mediterranean basin.

Appendices

Appendix 1

Example of a Chronological Timeline of Activities: Short Term of the Flood

This appendix refers to Chapter 2 on the social effects of flooding.

Legend

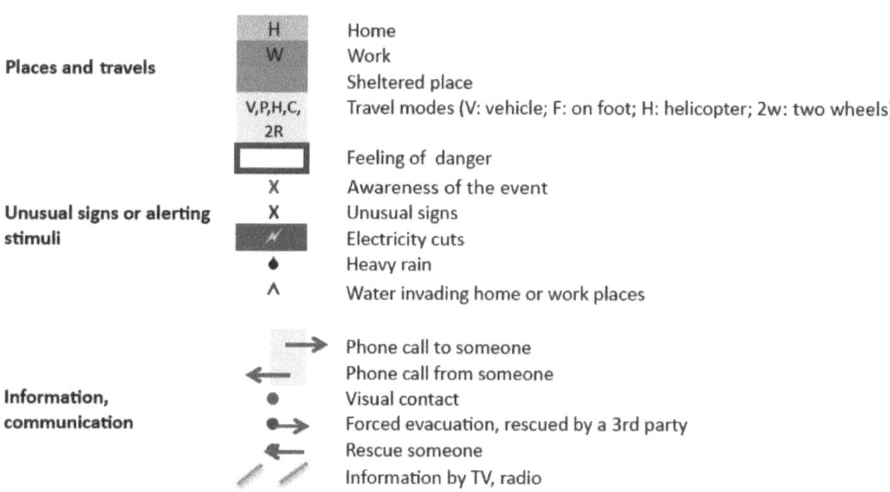

Chapter written by Isabelle RUIN.

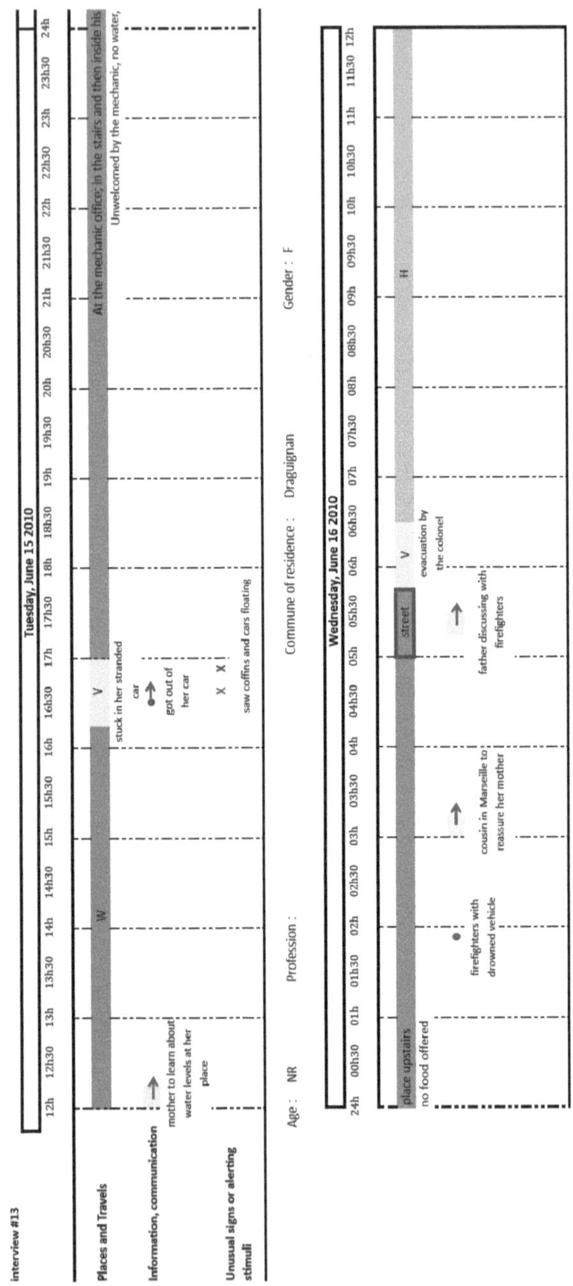

Figure A1.1. *The social effects of flooding. For a color version of this figure, see www.iste.co.uk/lutoff/hydrometeo1.zip*

Appendix 2

Methodology for the Socioethnographic Survey on "Living Practices"

A2.1. Introduction

This appendix refers to Chapter 2 and the social effects of flooding.

Contemplating the notion of "residential trajectories" for the analysis of inhabitant practices involves recognizing the fact that the position(s) filled by each individual changes over the course of time and that these successive positions are not due to hazard but follow on from each other in an understandable order [AUT 10]. In other words, it must be considered that mobilities have purpose. Our initial postulate is that this order depends as much on logic at the individual scale (professional, family and accidental, etc.) as on structural logic (regulations, etc.) [GRA 08] and natural logic (frequency and level of impact of events). The aim is not only to describe these trajectories but also to assign them reasons by providing an explanation that considers both the actors within the situation as well as structural and environmental determinants. The structural determinants are informed by the analyses presented in other chapters of this book (notably Chapters 5 and 6). The ethnographic multisite approach detailed in this part aims to expand upon the situation in these individual dimensions. Of course, the work between the different facets was interdependent: the different surveys and analyses necessitated constant intertask exchanges.

The investigation aims to remove the question of flood risk from inhabitants' daily lives, that is removing it as a dimension of a life considered in its entirety. The interest in observing ordinary practices constitutive of living in a risky environment is based on what these practices might teach us about taking (or not taking) the risky

Chapter written by Séverine DURAND.

dimension of the environment into consideration, and its variations at various times [DUR 14]. Moving away from the event raises the question of risk, not in itself but in daily life, and the study of the "risk public" in a "life with risk" [NOV 08]. The mode of living is understood here in the sense of a continuous practice, the fruit of individual and social experiences, and building a dynamic link with a space [STO 07]. It involves investigating the potential readjustments in modes of living (practical changes) for an environment that is deemed to be risky.

The study methodology used was socioethnographic. Residential trajectories were obtained through a narrative of inhabitant practices (section A2.3) in the context of a multisite ethnographic survey (section A2.2), applied in three towns of the project's study area[1]. These towns were selected for the contrasts they provided in terms of size, economic dynamics and exposure to risk. It was a question of moving from the particular to the general through the comparison of specific cases. The survey is envisaged over the long term (24 months into the survey in question here) with the aim of multiplying the survey temporalities (the practice of observation and carrying out interviews with individuals), around flood events when there are any but most of all, at a later time from these events, in the ordinary life of spaces.

The ethnosociological survey is implemented through case comparisons, ensuring that the comparisons remain plausible and pertinent in each unit of social object (living in a territory exposed to flooding), in the consistency of the initial set point (framing of individual interviews) and in the choice of individual cases for comparison (variations within a same social universe, i.e. rigorous sampling according to objectives). Sampling must be rigorous based on the survey objectives. Recourse to the classic, so-called *snowball method*[2] must carefully respect a certain balance in population categories. The first interviews, of an exploratory nature, are led with reference people in town or in connection to the theme (municipal team, associative representative, etc.). From these first meetings that are formally set up (initial contact by mail with a clear presentation of the project and methodology), some initial inhabitant contact details can be transmitted, and thus "the snowball is thrown". Next, to pursue the constitution of a corpus, it is necessary to vary the standard sociological categories (gender, age, activity type) of the surveys but also the categories that could potentially influence the type of story researched, such as for the present case: the localization (in or out of the flood area, according to the PPRI regulatory cartography), the occupation status (rented vs. owned), as well as the type of housing (town house, individual house with garden, apartment: ground level or first floor).

1 Department of Gard, project Mobiclimex (ANR 2013-2017), see Foreword.

2 A first interview provides the contact for the second, etc.

In order to attempt retracing the objectivity of pathways through a variety of discourses, the data obtained from the life histories must be made comparable. Before the moment of comparative analysis, it is the "comparative spirit" that permeates the entire survey [BER 97, p. 96].

A2.2. Analysis of the residential trajectories in flood-prone areas

The proposed protocol is characterized by a first phase of participant observation at the study sites (1), led in parallel to a follow-up of local contemporary preoccupations (2), within the analysis of the sociological, economic and political elements framing the situation (3).

A2.2.1. *Exploratory phase: permeate, appropriate, inquiring*

This first phase of the study consists of exploring the field and passing some time there by adopting a position of "naturalist observer" [JOS 03]. The first observation tool consists of repeated attendance time in town and roaming it to understand its nature to become permeated in the local atmosphere and glean information on the practices of local inhabitants. The repetition of these ambulatory practices generates many situations for leading a veritable "hunt for details", for information on living here as the different traces (discursive, practical, material) attest to the local awareness of the potential for flooding, or in contrast, its absence. This observation is carried out with the meticulous upkeep of a field notebook. The "ethnographic discipline" [WEB 91] then comes into play and allows the documentation of all the observations with minutiae and a sense of detail. To use Weber's words, this involves "documenting the observed or shared events and thus accumulating material for the analysis of the practices, discourses and interviewees positions" [WEB 91, p. 72]. This accumulation cannot be applied over the long term, with a regular presence in the field. Through repeated, supported and thorough observations at different time points and in different situations, the addition of these details takes on meaning in the eyes of the ethnographer. The observations documented in the field notebook take on meaning in their articulation, and then in terms of the other survey materials.

The second stage of the exploration work aims to obtain information on the practices of local inhabitants, that is, on what living here means. The main difficulty lies in the determination of observation locations. According to the configuration of urban life, central squares, cafes, stores, libraries, town halls, tourist offices and parks constitute the main resources. In these locations, it is as much a case of listening to the subjects discussed as it is collecting general information by encouraging informal exchanges with inhabitants (contacts who can then provide the resources for interview contacts). This involvement in local life then allows a more thorough understanding of the initial information. Through the participation in commercial (buying things in stores, weekly markets, etc.), festive (village fetes, various events, etc.) or associative activities (nature outings, workshops open to non-inhabitants, etc.), it is a case of varying trips to the field to approach inhabitants' ordinary life and multiply the opportunities for exchanges with them.

A2.2.2. *Follow-up of local preoccupation subjects: keeping watch on the local press*

In parallel to the observation time in the field, sustained attention is given to local media. The department's local radio station is regularly listened to collect information on local events, but also on the nature of the topics addressed. This regular listening of the radio is accompanied by an even more meticulous tracking of the written press. Readers of the local newspaper can form a "state of the public" in the sense given by Bordreuil [BOR 09] in reference to the works of Goffman[3]. This public forms a "tracking community", an "imaginary community", that of the press readership who "rely on the sense provided by the reading of a periodical to be in line with a split temporal pulse" [BOR 09]. This common reading defines both a common and a contemporaneity. By considering this "imaginary community" [BOR 09] that attention is given to the press, at several levels. First of all, particular attention is given, during the entire conducting of the fieldwork, to different news, whether local, national or even international, in terms of how they might affect the state of the local preoccupations on matters of flooding. It involves giving attention to the events, notably natural disasters or more specifically, floods that punctuate the news. Their effects (or lack of effects) enrich the other dimensions of observation.

To fine-tune this, a more thorough tracking of the local press is done to provide a view of the forces present and the significant actors in the local public arena [LEF 06]. Finally, and by further tightening the syntopic dimension, an even greater attention must be given to the commune's local newspaper, if it exists of course. The municipal paper covers the town's communication and its events. Beyond a follow-

3 In his work on the qualification of "public states", Bordreuil [BOR 09] distinguishes several levels of public that have the common point of connecting individuals who do not know each other. The first level (public state no. 1), in which Goffman's "Public Relations" documented their routines and rituals, refers to the established configurations between members that are unknown to each other but engaged in situations of co-presence. The second level (public state no. 2) refers to social configurations in which "unknowns" follow the same course of an event. They are no longer only "syntopic" (located in the same place) but also "synchronic" (situated in a same moment). This public is defined in how the individuals dedicate their attention, in parallel, on a same focus in a situation of co-presence, between strangers or not.

up that informs on the town news and events (providing as many entries as possible over the observation time), the analysis of the town's communications is interesting for several reasons. First of all, articles from a local commune newspaper reflect the discourse of the town hall, as it is written by appointed municipals or employees. Without denying the plurality within the institution, the contents of a local paper have the particularity of a politicized editorial approach, and in terms of a reflection of communal management, very little space is given for potential controversy [TIS 07]. More precisely, this specific type of press is interesting in how it reflects the role that the municipal team wants or thinks it should play on certain topics, and thus, through a mirror effect, how the local politicians perceive the expectations of its inhabitants. On the other hand, the local paper, by recounting the social, political and economic, etc. life of the living area, constitutes some of the paper press information available closest to the daily life of inhabitants. The underlying hypothesis is that this proximity participates in the creation of a common sense for inhabitants, a form of socialization through the construction of a specific public, a form of "imaginary community" [BOR 09], that of the communal newspaper readership[4].

A2.2.3. *Analysis of the structural elements (physical, economic and political) framing the situation*

In the context of this work, influenced by the urban ecology that considers an object as drawn by its occupation of the natural space where ecological and political dimensions are inextricably interlinked, the historicization of data is more specifically oriented toward the history of the town's urbanization. Without carrying out the work of a historian, this involves outlining the main lines in the history of the town and its population's development, at least for the contemporary period. For the earlier period, the work of historians remains the main resource. Moreover, this also involves studying the history of local risk management as well as the socioeconomic situation of the territory. Here, the study uses statistics as well as works developed through other branches of the project (ANR Mobiclimex, see Foreword) in which this study fits in.

A2.3. Analysis of narratives of inhabitant practices

> "A life story can constitute a remarkable instrument for the extraction of practical knowledge, as long as it refers to the description of experiences personally lived and the contexts within which they occurred" [BER 05, p. 23].

4 Without assuming that it is read by all the inhabitants, the ethnographic approach reveals, in successive fields, a regular mobilization of the newspaper in current discussions, leading to the supposition that it is followed by a majority of inhabitants.

Through the conducting of individual interviews, the experiences lived by inhabitants can be considered, attention can be given to the meaning given by the actors to their own behavior and experiences. By not focusing only on the floods, the interview is framed around the personal history in connection to his habitat.

From an ethnosociological perspective, the interviews aim to shape the inhabitants' narratives into stories of situational practice. To do so, the method of life stories is retained, focusing here on accounts of inhabitant practices. This conception assumes that there is a story as soon as the subject speaks of the experience lived. This narrative discursive production does not, however, exclude other forms of narrative (description, explanation, evaluation) that are integrated into the story and contribute to giving it significance [BER 15].

It is necessary to distinguish the history that we could call "real" from the narrative that comes of it. From the experiences lived by one person and their being told as a narrative, mediations undoubtedly intervene. Focusing on these mediations leads to the conclusion that each narrative is a subjective reconstruction. The study of these mediations, which may fall under the fields of psychology or linguistics for example, is not the object of this present study that addresses sociohistoric phenomena and in this sense, aims to overcome, among others, individual testimonies [BER 15].

> "It is their own experiences that individuals endeavor to describe [...]. The intervention of signaled mediations barely affects the diachronic structure of situations, event or actions that punctuated this life course. To use a metaphor, its 'image' is well restituted; however, remembering it may retrospectively modify the colors" [BER 15, p. 41].

Part of the knowledge that people have acquired through their direct experience with the situation (living in a territory exposed to flooding) must be extracted without looking too deeply into the singularity of each of these experiences. The different testimonies of the experiences lived in a same social situation must be connected – passing over the singularities – to reach, through the analysis of these narratives, a sociological representation of the (shared) social elements of the situation in question [BER 15]. In other words, a series of testimonies on the same social phenomenon – living in a region exposed to flooding – can be constituted and thus the common elements can be drawn and their social dimension revealed. It is precisely on the side of facts and practices (and not on that of representations) that these elements must be found.

Reducing the field of observation to a specific course (i.e. residential trajectories) and context (i.e. exposure of a territory to flooding) allows for a broader generalization that a study on the global formation of life trajectories would render impossible. This form of empirical data collection allows the follow-up of the formation of residential trajectories: it becomes possible to grasp which mechanisms and processes led individuals to find themselves in given situations (in this context, living in a flood-prone area) and how they manage this situation or attempt to remove themselves from it [BER 15].

A2.3.1. *Carrying out the interviews*

The individual interviews fall under so-called comprehensive sociology [KAU 96]. The comprehensive approach leans on the conviction that individuals are active producers of the social, and custodians of important knowledge that must be grasped from the inside. Nevertheless, sociological work does not limit itself to this statement: on the contrary, for the researcher it consists of being able to interpret and explain based on the data collected. The comprehension of the person is but an instrument: the aim of the sociologist is the comprehensive explanation of the social [KAU 96, p. 23].

It is a case of considering interviewees as people expressing their opinion, their experiences and their definitions of experiences lived. They can be led toward an explanation of a point of view by adopting the least prescriptive attitude possible. The main intention is to initiate and exchange through a conversational mode by attempting as much as possible to break the hierarchy of the interviewee–researcher interaction. The interview situation is seen as a situation of dialogue marked by confidence [DEM 97]. Through the tone used and the manner in which things are said, the importance of his point of view is indicated to the interlocutor, and the context of an ethnographic study becomes secondary to the encounter. Leading interviews at the interviewee's home – all the while observing the inhabitant practices: housing type, personal protection measures or not, etc. – contributes to facilitating an atmosphere of conversation. The frequency of personal level incursions during the interviews attests to a certain success in the willingness to "break the ice" and to make them analyze their position statements.

In the dialogical forms of narratives on practices, the inhabitant is invited to consider their experience through a filter: that of their residential trajectory. This framing, common to all the interviews, equates to form of "interview pact" that leads and frames the latter [BER 15]. For the entire corpus formed by the interviews to be considered as a whole, meaning that it feeds into a single common analysis, it is imperative that the initial context – the study subject designated by the researcher – be the same in all interviews, as it is here "Living in an area exposed to flooding".

Using semistructured interviews, the inhabitant should be led to give the narrative of their personal history in line with this subject. The person is invited to tell their story in connection to their housing, with a longitudinal perspective. Where were they born? Where did they grow up? How was the accommodation? Where did they go when they left their parents' home? What were the events that led them to perhaps move again? In each timeframe (successive homes) mentioned by the interviewee, it is necessary to then ensure that they clarify what sort of housing, the context, whether direct or more distant, what they appreciated, or in contrast, what the bad memories, etc., of this home were and the "portion of life" lived there. In this story, it is not a case of directing the inhabitant to the question of flooding but on the contrary, to "let them come to it", waiting for the opportune moment to follow up on the details and implications that this event may have had on their residential trajectories.

Moreover, it is an opportunity to question the inhabitants on their housing and their current daily practices. Does their housing suit them? What are the advantages and inconveniences? Do they wish to stay or to leave? Do they live there permanently? Are they often home? Do they work abroad? These elements allow the evaluation to feed into the work of other members of this research's broader project (ANR Mobiclimex, see Foreword), working on the vulnerability of people in their daily mobility. Finally, when the case arises, the experience of past floods is questioned and supports, in parallel, the works of the project team that this research on the moment of the "crisis" feeds into.

A2.3.2. *Analyzing the interviews*

The interviews do not provide facts, but words that rebuild what the person questioned has experienced (or is currently experiencing) and provides a view of the world defined at the individual level [DEM 97, p. 7]. Thus, these "worlds" that the interviews aim to grasp are not simply revealed. The mode of analytical treatment of the constituted corpus is just as important as the position adopted in its collection. The analysis of the corpus "of the words of the inhabitants", following a thematic analysis, aims to consider the "natural categorizations" (in opposition to scholarly or official categories) of the interviewees as the key process in the "constitution of sense in the discourse" [DEM 97, p. 9]. This analysis must first consist of completely retranscribing all the interviews, making sure to note the details concerning the interview situation (time of day, location, weather, how the interview was set up, etc.) in order to make the situation as clear as possible for later reading, over the course of the analysis. These transcriptions are carried out by scrupulously noting all the elements: the words of course, but also all the silences, hesitations, intonations, etc., that are as interesting in terms of gaining information as the words themselves, and are then also included in the analysis. Variations in the tone or volume of the voice are indicators of the emotional state or the interest generated by the discussion, and the silences or the unsaid carry sense in the same way as words. Langumier [LAN 06] showed in ethnographic interviews that he led after the 1999 flood incident in Aude for example, that silences and what was unsaid revealed the "indescribable for the ordinary moral", things that could not be expressed through words. On the basis of these integral retranscriptions of meticulous detail, the interviews are then the object of a thematic analysis, evolving on the basis of the "natural categories" that are revealed.

This thematic and comprehensive analysis of the interviews is completed with work on the factual elements, through the constitution of individual chronological timelines that permit a representation of the successive living locations in terms of the different events that carry a potential influence. Color codes allow the chronological representation of *personal events* (marriage, divorce, birth of children, etc.), *professional events* (education, finding work, loss of work, retirement) and *flood events* in terms of the changes in area of residence.

A2.4. Conclusion

The "moment" of the comparative analysis constitutes the core of the socioethnographic study [BER 15]. It is through the contrast of data collected from different inhabitants (i.e. "case") that a model can be elaborated – gradually refined over the course of case comparisons – for individual residential choices. From this analysis, the recurrences, shared elements, or on the contrary the contrasts depending on the context, must be found in order to reveal the specificities of living in a flood-prone area. This analysis is integrated and shaped around the other material collected through observation and the documentary analysis. The comparison of biographical pathways reveals recurrences and similar logics for actions. It is through this comparison work that the initial hypotheses, or those elaborated over the course of the comparative work, are clarified, confirmed and take on a wholly sociological form [BER 15]. Finally, the "negative cases" allow the consolidation or reformulation of these initial hypotheses. Thus, the moment is reached in which the saturation of information allows the study to be considered finished, at least provisionally.

By repositioning these elements of analysis in literature on residential trajectories obtained more broadly, it is possible to formulate, from this case study, hypotheses on residential trajectories in flood-prone areas.

A2.5. Bibliography

[AUT 10] AUTHIER J-Y., BONVALET C., LEVY J.-P. (eds.), *Élire Domicile. La construction sociale des choix résidentiels*, Presses universitaires de Lyon, Lyon, 2010.

[BER 97] BERTAUX D., *Le récit de vie*, Armand Colin, Paris, 1997.

[BER 15] BERNARD G., "Ruptures urbaines: une pragmatique spatio-temporelle", *Parcours Anthropologiques*, vol. 10, pp. 46–64, 2015.

[BOR 09] BORDREUIL S., "Société et configurations publiques de la vie sociale", *Séminaire les trames publiques du social ou Comment placer le public en société,* MMSH, Aix-en-Provence, 2009.

[DEM 97] DEMAZIERE D., DUBAR C., *Analyser les entretiens biographiques,* Nathan, Paris, 1997.

[DEW 95] DEWEY J., *Logique. La théorie de l'enquête,* Presses universitaires de France, Paris, 1995 (1938).

[DUR 14] DURAND S., Vivre avec la possibilité d'une inondation? Ethnographie de l'habiter en milieu exposé... et prisé, PhD Thesis, Université Aix-Marseille, 2014.

[GRA 08] GRAFMEYER Y., AUTHIER J.-Y., *Sociologie urbaine*, Armand Colin, Paris, 2008.

[JOS 03] JOSEPH I., *Erving Goffman et la microsociologie,* Presses universitaires de France, Paris, 2003 (1998).

[KAU 96] KAUFMAN J.C., *L'entretien compréhensif,* Nathan, Paris, 1996.

[LAN 06] LANGUMIER J., Survivre à la catastrophe: Paroles et récits d'un territoire inondé, PhD Thesis, EHESS, Paris, 2006.

[LEF 06] LEFRANC C., TORRE A., "Les conflits dans les zones rurales et périurbaines. Premières analyses de la presse quotidienne régionale", *Espaces et sociétés,* vol. 2–3, nos. 124–125, pp. 93–110, 2006.

[NOV 08] NOVEMBER V., PENELAS M., VIOT P., « 'L'effet lully' : un territoire à l'épreuve d'une inondation », *Cosmopolitiques*, no. 17, pp. 89-106, 2008.

[STO 07] STOCK M., "Théorie de l'habiter. Questionnements", in PAQUOT T., LUSSAULT M., YOUNES C. (eds.), *Habiter, le propre de l'humain. Villes, territoires et philosophie*, La Découverte, Paris, pp. 103–125, 2007.

[TIS 07] TISSOT S., "De l'emblème au problème: le destin des grands ensembles dans une ville communiste", in TISSOT S., *L'État et les quartiers. Genèse d'une catégorie de l'action publique,* Seuil, Paris, pp. 147–185, 2007.

[QUE 05] QUERE L., OGIEN A. (eds.), *Le vocabulaire de la sociologie de l'action*, Ellipses, Paris, 2005.

[WEB 91] WEBER F., "L'enquête, la recherche et l'intime ou pourquoi censurer son journal de terrain?", *Espaces Temps*, vol. 47–48, pp. 71–81, 1991.

Appendix 3

Field of Study of the Socioethnographic Survey on Living Practices

This appendix refers to Chapter 2 on the social effects of flooding.

The three study sites for the socioethnographic survey were the localities of Collias, Sommières and Le Vigan (department[1] of Gard, France).

Collias is a village of 1,112 inhabitants[2], located close to the famous *Pont du Gard*, in the east of the department. A small rural village, which was historically agricultural and touristic, Collias lies close to main circulation routes and becomes increasingly attractive for the periurbanization of urban centers such as Nîmes and even Avignon. During the 2002 flood crisis that affected nearly the whole department, several houses were carried away and many inhabitants had to be relocated. Around a hundred homes were heavily impacted by the floods, some completely carried away by the river. Faced with such extensive damage, the State (supported by local authorities) fell back on the Barnier funds[3] for the expropriation

---

Chapter written by Séverine DURAND.

1 In the administrative divisions of France, the department is one of the three levels of government below the national level ("territorial collectives") between the region and the commune.

2 Population census INSEE 2014.

3 The major natural risk prevention fund (FPRNM, also called the "Barnier fund") was created by law no. 95-101 from February 2, 1995, relative to the reinforcement of environmental protection. This fund was initially destined to finance compensation for the expropriation of property exposed to a major natural risk. Its utilization was then expanded to other categories of expense. The current scope of action of measures eligible for the funds is defined in article L561-3 of the Environment Code. The Barnier fund is fed by a 12% levy on the "natural disasters" premium for housing and automobile insurance contracts. (Source: *www.eure.gouv.fr*).

of 90 inhabitants (48 main houses, 42 second homes). The affected inhabitants were offered sites for construction at reduced prices in an area of the commune that was not prone to flooding. It is this specificity that makes this site of interest for the survey, by presenting a sample of families that have been relocated.

Sommières is a town of 4,644[4] inhabitants. This medieval town has the particularity of having been built in the riverbed. With a very rich history, marked by silk manufacture and trade, for centuries, Sommières has benefitted from a constant attractiveness. Despite regular floods that affect it, today it still remains attractive due to its tourist interest, but also due to its proximity to the main roads leading to Montpellier and Nîmes, the two nearby urban centers. This town is interesting because it cannot be further protected against floods: they will always occur.

Le Vigan is a town of 3,933[5] inhabitants located in a rural valley, in the north-west of the Gard department. Today, the town is in economic decline, although it shined for centuries. Like many towns in the Cévennes, it owes a large part of its prosperity to the hosier industry (silk spinning and stocking knitting). The traders' greatest richness was in wool work and sericulture, through which stockings could be knitted to be exported all around Europe. The town was the administrative capital of Western Cévennes, important due to its large markets and its role as a landmark town at the foot of the Cévennes. The textile industry flourished there in the 20th Century. More recently, several business relocations have affected the local job market. This town is of interest due to its difficult economic situation. In contrast to the other sampled towns, census data shows that the population is in substantial decrease here.

4 Population census INSEE 2014.

5 See footnote 4.

Appendix 4

Example of a Residential Chronological Timeline – Long Timeframe of Life Trajectories

This appendix refers to Chapter 2 and the social effects of flooding.

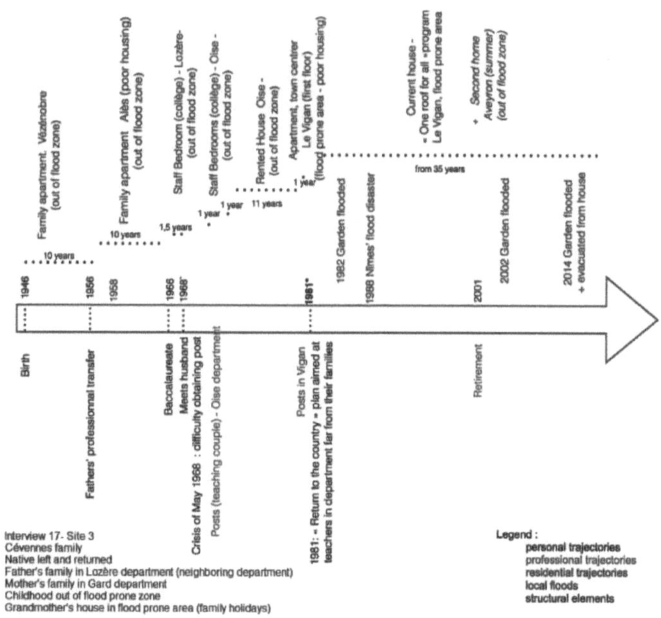

Figure A4.1. *The social effects of flooding. For a color version of this figure, see www.iste.co.uk/lutoff/hydrometeo1.zip*

Chapter written by Séverine DURAND.

Appendix 5

Methodology for the Analysis of Migratory Dynamics

This appendix refers to Chapter 7 and the analysis of residential migrations.

Among the justifications for using modeling and simulation is that of rapidly recognizing an incomplete understanding of the observed field due to heterogeneous and dispersed information. Computer simulation is an effective method for measuring the importance of factors identified in their spatial and temporal dimension, and for completing this understanding by orienting supplementary research to improve the quality of the model.

To analyze the migratory dynamics within Gard, we decided to build a mathematical (multiple regression) model, applied differently for each commune based on the number of explanatory factors identified and the weighting selected for each intercensal period. The model was made using the GAMA platform in order to visualize the model's results and to measure, for each increment of time, the discrepancy accumulated compared with the reality. For the present case, we gradually built the model, starting with the factors that were applicable to a large number of communes, in order to move toward more local specificities revealed through the poor results obtained in the simulations. The discrepancies with reality guided the literature search to identify the factors at the origin of the observed migratory specificities.

For each commune, we therefore built a multiple regression integrating the different indices whose weighting varied depending on two elements:

Chapter written by Edwige DUBOS-PAILLARD, Antonin PAVARD and Joël BOULIER.

– the intercensal period for taking the observed demographic evolution into account, notably the decrease in average annual variation rates between 1968 and 1975 (Figure 7.1);

– the different types of spaces identified either through the initial literature review or through the discrepancies observed on maps from different simulations.

Moreover, considering that a portion of migrations are not explicable through the factors identified in the model, we added statistical noise. The latter, of very weak value, is determined via the following formula:

$$\text{Statistical noise} = 0.05 \times (1 - \text{random}^{1}(2{,}000)/1{,}000)). \qquad [A5.1]$$

The evolution of the population linked to theoretical migratory rates within the different communes is found using the following formula:

$$\text{Theoretical population } N + 1 <= \text{Theoretical population } N + (\text{theoretical population } N \times \text{coefficient of variation for the theoretical annual migration rate}) + \text{natural annual rate observed.} \qquad [A5.2]$$

Given that for each intercensal period:

$$\text{Coefficient of variation for the theoretical annual migratory rate} <= ((1 + \text{noise}) \times \sum (\text{index}_n \times \text{weight of index}_n \text{ current year})/\sum (\text{weights of index}_n \text{ current year}) \qquad [A5.3]$$

The reference population is that of 1968. For each year, we added the theoretical migratory rate and the natural rates observed. Next, we compared the theoretical population with the communal population observed in 1975, 1982, 1990, 1999 and 2010.

The calibration of the different indices selected for each intercensal period was done in an iterative manner. Different parameters were tested in order to get as close to possible to reality. As such, the periurban communes of Nîmes have a different weighting to those of Alès, even though the model used is the same: that of differential urbanization [VAN 82].

Bibliography

[VAN 82] VAN DEN BERG L., DREWETT R., KLAASEN L. *et al.*, *Urban Europe: a Study of Growth and Decline*, vol. 1, Pergamon Press, Oxford, 1982.

1 Random: random sampling of a value.

List of Authors

Sandrine ANQUETIN
Université Grenoble Alpes
CNRS, IRD, Grenoble-INP
Institut des Géosciences de l'Environnement
France

Marco BORGA
Department of Land, Environment, Agriculture and Forestry
University of Padua
Italy

Joël BOULIER
UMR 8504 Géographie-cités
Panthéon-Sorbonne University
France

Isabelle BRAUD
Irstea Hydrology–Hydraulics Research Unit
Lyon-Villeurbanne Center
France

Jean-Stéphane CARNEL
GRESEC
Université Grenoble Alpes
France

Jean-Dominique CREUTIN
Université Grenoble Alpes
CNRS, IRD, Grenoble-INP
Institut des Géosciences de l'Environnement
France

Edwige DUBOS-PAILLARD
UMR 8504 Géographie-cités
Panthéon-Sorbonne University
France

Véronique DUCROCQ
CNRM-GAME/MICADO
Météo France, CNRS
Toulouse
France

Séverine DURAND
UMR PACTE 5194
Université Grenoble Alpes
France

Sylvie DUVILLARD
UMR PACTE 5194
Université Grenoble Alpes
France

Benoit LAFON
GRESEC
Université Grenoble Alpes
France

Céline LUTOFF
UMR PACTE 5194
Université Grenoble Alpes
France

Gilles MOLINIÉ
Université Grenoble Alpes
CNRS, IRD, Grenoble-INP
Institut des Géosciences de l'Environnement
France

Antonin PAVARD
UMR 8504 Géographie-cités
Panthéon-Sorbonne University
France

Isabelle RUIN
Université Grenoble Alpes
CNRS, IRD, Grenoble-INP
Institut des Géosciences de l'Environnement
France

Valérie SANSEVERINO-GODFRIN
Mines ParisTech
Sophia Antipolis
France

Béatrice VINCENDON
CNRM-GAME/MICADO
Météo France, CNRS
Toulouse
France

Bruno WILHELM
Université Grenoble Alpes
CNRS, IRD, Grenoble-INP
Institut des Géosciences de l'Environnement
France

Index

A, C

D, E, F

G, H, I

L, M, N

P, Q

R, S

T, U, V, W

FSC
www.fsc.org
MIX
Papier | Fördert gute Waldnutzung
FSC® C083411

Kolibri 360 GmbH
Ferdinand-Jühlke-Straße 7
99095 Erfurt, Deutschland
produktsicherheit@kolibri360.de